Ethics for Engineers:
A Brief Introduction

Aristotle 384-322BC (Roman copy in marble of a Greek bronze <u>bust</u> of Aristotle by <u>Lysippos</u>, <u>c</u>. 330 BC)

Immanuel Kant 1724-1804 (painter unknown) Reproduced under Creative Commons licence (CC BY-NC-SA 4.0)

Ethics for Engineers: A Brief Introduction

Anthony F. Bainbridge

CRC Press
Taylor & Francis Group
Boca Raton London New York

CRC Press is an imprint of the
Taylor & Francis Group, an **informa** business

First edition published 2022
by CRC Press
6000 Broken Sound Parkway NW, Suite 300, Boca Raton, FL 33487-2742

and by CRC Press
2 Park Square, Milton Park, Abingdon, Oxon, OX14 4RN

© 2022 Anthony F. Bainbridge

CRC Press is an imprint of Taylor & Francis Group, LLC

ISBN: 9781032076904 (hbk)
ISBN: 9781032077086 (pbk)
ISBN: 9781003208433 (ebk)

DOI: 10.1201/9781003208433

Typeset in Times
by Deanta Global Publishing Services Chennai India

DEDICATION

For my Grandson Leo
who will surely become more ethically aware
than his Grandfather has been

The most important human endeavour is the striving for morality in our actions. Our inner balance and even our very existence depend on it. Only morality in our actions can give beauty and dignity to life.

Albert Einstein

It is clear, then, …. that it is not possible to be good in the strict sense without practical wisdom, or practically wise without moral virtue.

Aristotle, The Nicomachean Ethics, vi, 13

Our whole life is startlingly moral. There is never an instant's truce between virtue and vice.

Henry David Thoreau

If to do were as easy as to know what were good to do, chapels had been churches, and poor men's cottages princes' palaces. It is a good divine that follows his own instructions: I can easier teach twenty what were good to be done, than be one of the twenty to follow mine own teaching.

Shakespeare, Merchant of Venice, 1, ii.

CONTENTS

PREFACE

This book began as a short series of lectures given to second-year engineering students in the Faculty of Engineering & Design at the University of Bath, as part of a Semester-2 course called Group Design & Professional Engineering Practice. The initial requirement was for a brief programme to introduce students to the statutory and regulatory aspects of engineering today and arose from a comment by an academic accreditation panel from the Institution of Engineering and Technology.

It became clear as the course developed, and as concerns grew in the world outside the classroom about the ethical basis of decision-making in business, that the course should draw attention to ethical concerns in product design. In any event it was also clear that the major issues with which I had to deal in the context of legislation, statute, and regulation raised fundamental ethical questions for the engineering profession – outstandingly so in the matter of product safety.

At about the same time the UK Engineering Council, the Royal Academy of Engineering, and all the major professional bodies promulgated guidelines on the significance of ethical awareness in the world of practising engineers. The Engineering Council UK (EC UK) has published a 4th edition of UKSPEC, the standard which maps the competences registered engineers were expected to demonstrate, incorporating a clear requirement for registration candidates to have thought about the ethical practices around them and to be ready to observe, comment on, and, if necessary, amend their ethical behaviours.

The aim and content of the course then became: 'To provide students with knowledge of the ethical and statutory frameworks under which professional engineers operate.' The course has run for over ten years, developing and broadening in the process. The task I always impose on myself never to teach the same course twice requires me to review constantly both my material and my delivery. Duty is a hard taskmistress.

This short book is not intended as anything more than a primer covering the main issues which inevitably arise in any discussion of ethics in the engineering context – or indeed in any context involving the applied sciences. The reference list is full enough to lead interested readers into deeper analyses and wider applications. My context is engineering for the simple reason that I am an engineer with a long career which has enabled me to touch on many fields within the engineering profession. I have been proud to call the IET – formerly the IEE – my professional home for 60 years.

The reader will find many rhetorical questions throughout this work, for which I make no apology. Any account of the role of ethical attitudes in a professional context faces the massive issue that in the end we decide for ourselves. The matter cannot be treated like a topic in an engineering manual, where in general one expects to find clear pointers and recommendations. Instead we must find our own answers.

I have occasionally followed the established usage 'he' and 'his' in referring to 'the engineer' in general. The welcome rise in the number of women engineers is already transforming the profession, and this process will surely continue; but I am uncomfortable with the use of 'they' to cover single members of either gender, and I find s/he clumsy. I have tried to work around the problem, but may not always have succeeded in doing so.

(Dr) Anthony F. Bainbridge
Part-time Teaching Fellow, University of Bath
April 2021

ACKNOWLEDGEMENTS

I am grateful to Dr Adrian Evans, Head of Department, and to my colleagues, Dr Francis Robinson and Dr Steve Pennock of the Department of Electrical and Electronic Engineering at the University of Bath, who, stepping into the unknown, invited me to create and deliver the course in 2011; for supporting me in my role and for substantial help, advice, and tolerance as I found my way around the departmental and university processes and procedures. Their welcome has enabled me to pace the lecture theatres of the University of Bath for many years. I have of course learned much from the insights of many students expressed after lectures or in their assignment reports.

My contacts with IET members and staff during thirty years of active involvement as a volunteer, primarily as Professional Registration Adviser and as PR Interviewer, have been of immeasurable value, for which I remain thankful. I am sure I have gained more than I have given.

The publication process has been led by Nicola Sharpe and Nishant Bhagat of Taylor & Francis, who have shown exemplary patience as I learned how the process works. The index has been prepared by Catherine Hookway of the Society of Indexers, to whom I am most grateful for coming to my rescue.

AUTHOR BIOGRAPHY

Dr Anthony F. Bainbridge MA CEng FIET FHEA
Anthony F. Bainbridge is a Teaching Fellow at the University of Bath in the UK. Educated at the University of Cambridge, he then worked in the aerospace industry for many years before beginning a new career as business consultant specialising in technology development and management; he is interested in the development of leadership competences in engineering and technology enterprises. He has delivered training courses to a wide variety of clients and to university undergraduate and postgraduate students. As a Fellow of the UK Institution of Engineering & Technology, he is an active volunteer; in the 1990s he founded and chaired the IET's consultancy network, and now mentors and interviews potential CEng registrants. He has published papers on 'Activity-based Costing', 'Securing Venture Capital', 'Decision-making for Innovation', and 'Is Management a Profession?' in the *IET Engineering Management Journal* and elsewhere.

He is married and lives in Wiltshire, England.

ETHICS IN THE PROFESSIONAL CONTEXT

Two immediate questions raised by any discussion of engineering as a profession, and by the need to consider ethical attitudes in the practice of that profession, must be: 'What does "professional" mean?' and 'What are "ethics"?'

It has not always been obvious that the two are closely linked. We assume that engineers, like professionals in other walks of life, have always behaved professionally, in accordance with their own understanding of the meaning of the word, but we know of many histories which reveal personal failings. It is now acknowledged that professionalism and ethical behaviour go hand in hand – see Davis (1991) for a wise overview on this issue. Many engineering courses refer to ethical standards; many employers require their employees to conform to corporate ethical charters; the approach to professionalism through registration to the United Kingdom Standard for Professional Engineering Competence (UKSPEC) explicitly requires candidates to have thought about ethics and to be aware of published standards.

We will briefly review ethical theory and our responses to it in Sections 2.1 and 2.2; here we must consider the meaning of professionalism and its bearing upon the work and attitudes of engineers. What is implied when we call ourselves 'professional' engineers? How are the commercial pressures of the business environment to be weighed against the need to treat all people fairly and act with integrity? What are the achievable limits on action in specific contexts?

The word 'professional' implies at least (a) formal acceptance by a peer group, and (b) conformance with published codes of practice. All professional engineers are aware of the many influences – practical, social, legal – which impact upon their work and are relevant to the design task. Against this backdrop engineers know how to convert a customer requirement into a design specification and produce a design to meet this specification. We now realise that ethical considerations

DOI: 10.1201/9781003208433-1

and legislation also impose their disciplines, and these must be woven into the design process from its inception.

The values by which we direct our professional behaviour matter to the public, to our employers, and to government; but why do they matter? There is, to be sure, a broad debate over personal and moral values in general; the perception is often expressed that that moral standards throughout Western culture are eroding. Whether or not this is true, the debate continues and becomes ever more pressing. Engineers must display the responsibility to scrutinise the ways in which their moral values impact on their professional work. The education and training of engineers in the 1960s did not emphasise ethical issues; it was not suggested at that time that the career of a professional engineer and manager might uncover ethical problems and dilemmas; if they did the engineer no doubt had to resolve them alone.

Now we know better. Many employers, perhaps the majority, publish codes of practice to be followed by employees individually and collectively and declare in so doing their commitment at the corporate level. Furthermore, professional institutions in the engineering field expect their members to conform to published codes of practice (see Annex 1 for the IET Rules of Conduct as typical examples). Two specific sections in the Bye-laws of IET – which are typical of many similar published guidelines – state:

> 29. Persons in any category of membership shall at all times uphold the dignity and reputation of their profession, act with fairness and integrity towards everyone with whom their work is connected, and towards other members and safeguard the public interest in matters of health, safety, the environment and otherwise.
> 30. Persons in any category of membership shall, in addition, exercise their professional skill and judgement to the best of their ability and discharge their professional responsibilities with integrity. They shall encourage the vocational progress of those in their professional charge.

Furthermore, many engineers wish to become registered as Chartered, Incorporated, or Technician Engineers in due course through many of the UK's professional engineering institutions. The relevant standards are set out in the Engineering Council's publication 'United Kingdom Standard for Professional Engineering Competence' (UKSPEC – see Annex 2) requiring candidates to be able to show evidence of the application of a whole spectrum of competences, including an awareness of ethical thinking and practice.

There is perhaps one further criterion to distinguish a profession from other groupings: we expect professionals to serve the public good. The American sociologist Talcott Parsons suggested many years ago that 'the professions have become the most important single component in the structure of modern societies' – a challenging comment even if you regard it as somewhat overstated. In an increasingly complex society do we need the professions more than ever in helping us to negotiate the world and to serve 'the public good'? Or conversely, at a time of much emphasis on empowered investors, stakeholders, and citizens, is the need for professions diminishing?

Traditionally there has been asymmetry of information between the knowledge which specialists have and that accessible to and understood by the public, but as the general public becomes more knowledgeable this gap is palpably narrowing. If nowhere else, we can see this in the medical profession where greater access to information has made patients more aware. The professions are being challenged to modernise themselves to adapt to these new realities; in this changing perspective how do we understand what the public good might be and how it can be served?

The Royal Society of Arts in the UK organised in 2002 a project 'Can the Professions Survive? Exploring professional values for the 21st century' which sought to establish a new model of a 'profession' combining the rigour of the traditional professions with the vigour of a new 'professionalism', encouraging professionals in the UK to become a more significant, trusted, and creative force for economic and social good. The project asked what society now requires from professionals and from professional practice; the intention was to revitalise attitudes and practice and to encourage rigour and excellence in the professions. It is not clear whether the RSA project achieved its full aims, but it must be said that the objectives were sound: Traditional professional bodies, and their membership in responsible roles, must be prepared to question their own rationale and to innovate as they seek solutions to the problems and challenges societies face anew with each new generation.

The word 'responsible' carries much weight. Responsibilities may be passive or active. They may be incurred through one's defined role, through the personal morality we bring to it, and through one's sense of professionalism applied in the work. Michael Davis (Davis 2012) identifies nine related senses of 'responsibility' – indeed, the anecdote which gives his paper its title models a pointless attempt to avoid responsibility.

How far does one's responsibility reach? It might be claimed that responsibility was restricted because of ignorance – the engineer did not know (perhaps had not been told) what he was responsible for; or he was not free to do what was required of him; or he did not have time to satisfy himself as to the right course of action, so could not approve it. Each of these may give rise to uncertainty as to how to behave; and if so, the professional engineer will seek advice – just as we would in cases where we became aware of gaps in our technical knowledge.

Responsibilities may, and often do, conflict. In the Challenger disaster (Figure 1.1) of 1986 (see, for example, the report of the Committee on Science & Technology of the US Congress (Committee on Science and Technology, 1986)), we can observe a very clear conflict between the responsibilities of the Morton Thiokol engineers to report their deep concerns over safe launch, and those of NASA senior management for whom political and public pressures dominated. Annex 3.8 expands on the Challenger case.

The engineer might also wish to claim that he could not be held responsible for decisions made or actions taken by an organisation to which he belonged. In the context of the present discussion, this organisation might be the employer or the professional institution. When the individual agrees to behave (always 'to the best of his ability') in ways mandated by such organisations, what corresponding duties are incumbent upon the organisation in return for his membership fee? He has no means of imposing reciprocal duties on the larger group; even in extremis withholding dues achieves nothing. W. H. Walsh asks in 'Pride, Shame and Responsibility' (Walsh, 1970): 'In virtue of my membership in some larger whole or wholes, how can

Figure 1.1 The Challenger Disaster 1986 – see Annex 6.

I reasonably be expected to take responsibility for what these bodies do in circumstances where I could have no conceivable influence on their actions?'

We should consider also the balance of responsibilities between the employer and the employee. Those of the employer may be expressed in terms of the idea of Corporate Social Responsibility (CSR). In the conduct of business, should organisations think about their responsibilities to employees, the wider society, and the environment? Notions like these are widely propagated, and their underlying principles are hard to evade. A business does not exist in isolation, nor is it simply a way of making money – although that is ultimately what it is for. How should the engineer act so as to ensure that outputs conform to, and extend, the corporate responsibilities of the organisation?

Employees, customers, suppliers, and the local community are affected by business activities. Products, and the way they are made, have an impact on the environment. The processes used by businesses may also have adverse impacts. Many statutory requirements impose duties on businesses; a sense of the social responsibilities of the business implies understanding the business' impact on the wider world and considering how this impact might be used positively. CSR implies taking a socially responsible attitude, going beyond the minimum legal requirements, following straightforward principles that can be applied to all businesses, whatever their size or sector. And in terms of the chief 'purpose' of any business, CSR may also be good for the 'bottom line.' A concern for CSR may lead an organisation to have regard to, for example, product design for reduced energy consumption, building construction for the same, sensitivity to the needs of the local community, renewed focus on training and education of employees, awareness of waste and scrap materials, healthcare for employees, and perhaps other issues. Corporate ethical charters frequently address these and other matters, and employees are expected to accept and practise the implications in all their work (Schwab, 1996).

There is another and seemingly contrary view. Milton Friedman (1970) suggests that the social responsibility of business is to increase its profits, and that all the other things it may wish to do are secondary, and are only possible to the extent that they are able to satisfy that first responsibility. Friedman was attempting to debunk the notion that corporations (meaning the executives and managers who work for them) have a responsibility to act to increase social welfare, regardless of the law on the issue at hand (whether that issue be discrimination, pollution, or something else). In his view, executives have a moral responsibility to act always in the long-run best interest of the shareholders. He is not arguing that businesses should never engage in activities that increase social welfare; in fact, he argues that free-market capitalism itself increases social welfare. He also notes that businesses certainly will engage in activities that result in increased social welfare. However, in Friedman's view the executive's sole motivation for such engagement must always remain long-term increase in shareholder wealth.

Elaine Sternberg (2000) supports Friedman and asserts that the specific and sole business objective is to maximise owner value over the long term by selling goods or services; she observes that while actual commercial enterprises often do much else, all other aims are subsidiary. On this view the principles of business ethics are those presupposed by the essential activity of maximising long-term owner value; such views call for confidence in a future, and hence for trust. Business must therefore be conducted with honesty, fairness, the absence of physical violence and coercion, and a presumption in favour of legality; but business is ethical only if it supports causes which maximise long-term owner value. This does not mean that a business cannot or should not behave in ways that are ordinarily considered to be socially responsible, such as hiring physically-impaired employees and supporting local schools; to the extent that such actions help maximise long-term owner value, they are wholly compatible with business.

For Sternberg, therefore, the whole theory of the Firm must have at its root an ethical basis. She offers an 'ethical decision model' to provide clarity in addressing ethical quandaries faced by the business. Certainly, in all such discussions, the engineer will recall that design and all other engineering activities take place within the business and must therefore be shaped by its ethical policies. It is, however, debatable to what extent the types of ethical problems faced by engineers are amenable to the decision model suggested by Sternberg, which concludes always by proposing a single 'right course of action.' Many recent substantial ethical conflicts, on the other hand – for example, those outlined in the case studies of Annex 3 – have been exacerbated by the sheer complexity of inter-company contractual relationships, conflicts of interest during the design process, multiple and complex causes, international influences, and so on. Such complexity prohibits the search for simple, single responses.

The perspective that a business's primary responsibility is to its owners rather than to society has generated, and will no doubt continue to generate, much controversy and debate (see especially Klein, 2006; Dunn 2006). It might be said that our attitude to ethical issues in business and in engineering may be determined in part by the line we take on Friedman. It is essential that sooner or later every engineer decide whether Friedman's perspective is appropriate or even useful, and think about its bearing on the individual's personal ethical code. At the very least this implies that engineers should seek to understand the business aims of the employer in framing their personal code in response. It is, after all, easier to promulgate a corporate system of applied ethics than to ensure that in each and every case its principles are followed. See Klein (2006) and Sternberg (1999, 2000) for excellent discussions of the purpose of business and the contrasting perspectives of stakeholders and stockholders.

However, a study by List and co-worker (List and Momeni, 2017), based upon a survey of 3000 employees in a number of firms, produced clear evidence, supported by other complementary treatments, to suggest that 'moral licensing' may be at work and that the 'doing good' nature of CSR induces workers to misbehave on other dimensions that hurt the firm. Some employees may act detrimentally towards the firm by shirking on their primary job duty when CSR was introduced. This survey serves to warn that business practices such as CSR should not be applied without careful monitoring of their impact.

Nissenbaum (1996) draws on philosophical analyses of moral blame and responsibility to identify four barriers to the proper identification of accountability in an ever more computer-dependent society.

THE ETHICAL BASIS FOR ENGINEERING ACTIVITIES

2.1 INTRODUCTION TO ETHICAL THEORY

We shall be concerned in this book with ethical principles and with the relevant statutory requirements in the context of professional engineering practice.

There is widespread agreement that engineers should be introduced at an early stage in their careers to the need for ethical thinking in preparation for their professional lives, both to encourage clarity about ethical issues and the practices within which they arise, and to develop skills in communication, reasoning, and reflection. Engineers will thereby better understand the nature of professional responsibility, be able to identify the ethical elements in decisions and to address and perhaps resolve problems; they will develop their critical thinking skills and professional judgement, understand practical difficulties of bringing about change, and develop an ethical identity to carry forward in their working life. The case studies outlined in Annex 3 illustrate the ways in which ethical conflicts have arisen in some real and difficult situations; they invite readers to assess some of the options facing the protagonists and formulate their own reactions.

If the phrase 'perhaps resolve' used above seems tentative, this is because it is likely that no ethically problematic context will repeat; every situation will to some extent present new features which render the automatic application of previous solutions questionable. Case studies may suggest possible approaches, as will earlier personal experiences, but there will often be contextual and personality differences which will call for innovation.

Ethical constraints may show up in many aspects of the engineer's work. For example:

- Safety: What does safety mean? Is there agreement on how far we can go to achieve it, and who decides?

DOI: 10.1201/9781003208433-2

- IPR: Who owns the rights to a design? How can we be certain that we do not unlawfully exploit protected IP?
- The environment: What impact will the design have on the environment? What costs may be incurred in addressing the problem?
- Boundaries: How can the engineer avoid being swept up by events hitherto unforeseen without going outside acceptable boundaries of knowledge or ethical behaviour?
- Quality: How is quality to be defined and be continually improved within the tight bounds of a fixed-price contract?

Given these constraints, who is responsible and accountable for design conformance? Where does the boundary lie between the responsibilities of the designer, the manager, the manufacturer, the end user? In this book, we consider some of the statutory requirements which bear upon the work of the professional engineer, and consider some of our ethical responses to them. These responses may spring initially from our upbringing, and later from our education and training; but there will come a point when we have moved from *head* knowledge: 'I *think* I know how to act' – to *heart* knowledge: 'I'm confident that this is how I always act.'

There seems no end to the publishing of books and journal papers on ethical theory and its applications in various contexts. It is not to be expected that engineers will find time to read more than a few of them, particularly at the stage of their careers when engineering studies absorb much of their attention. But even a short course in applied ethics cannot evade the need to direct the student towards a selection of the texts most relevant and approachable. Students with deeper interest may find time to go back to the first and greatest – and still relevant – thinkers, Aristotle and Immanuel Kant; it remains true that all conversations on ethical theory lead back sooner or later to Aristotle's 'Virtue Ethics' in *The Nicomachian Ethics* (Aristotle, 2009) and to Immanuel Kant's deontological ethical theory published in his *Groundwork on the Metaphysic of Morals* (see especially Walsh, 1975 and Körner, 1990). These, the two foremost philosophers to have thought and written about ethics, remain worthwhile, if challenging, reading in the many editions available. The reference section lists those sources ancient and modern which seem to speak most clearly to interested professionals.

We must start by defining our terms. Here again we face a difficulty – many of the major writers on ethical theory have created their own definitions. We begin with a definition which seems to me closest to the mark:

Ethics is the discipline that examines our moral standards, and the moral standards of our society, to evaluate reasonableness and its implications for our lives and decisions - standards of personal and collective behaviour which illustrate our values and to which we are expected to aspire as members of a civil society.

Our values (or morals) are the personal standards by which we assess significance in matters of right and wrong (of attitudes), or of good and bad (of actions).

These ideas can be expressed in many contrasting ways, each of which seeks to convey elements of the full picture:

- In essence, ethics is concerned with clarifying what constitutes human welfare and the kind of conduct necessary to promote it (Powers and Vogel, 1980).
- Ethics is knowing the difference between *what you have a right to do* and *what it is right to do.*
- Ethical theory as a branch of philosophy seeks a reasoned examination of what custom tells us about how we ought to live.
- Morality you believe – ethics you learn.

The idea that ethics is a *discipline* (as in the first definition above) implies the need for constant re-examination of our attitudes, rather than a once-only commitment. Our ethical positions may change throughout our lives – it would be perhaps surprising if we did not modify them as fresh experiences influence us – but always there is the sense of a commitment to a code which we have consciously adopted, rather than unthinking adoption of the code of the social group into which we were born or recruited.

It should be clear that the underlying truths are common to all perspectives. But is engineering as a profession 'normative' – i.e. does engineering tell us how we *ought* to do things, or rather does our upbringing and our working and living environment communicate these signals? Whence do we get our sense of the difference between 'I will do ...' and 'I ought to do ...'? These questions are beyond the scope of this short guide, but should be addressed by anyone seeking to establish a secure basis for ethical behaviours. Words such as *good, bad, right, wrong,* and *ought* must underpin any thinking of the ethical basis on which we base our decisions and actions.

A good starting point is to consider the sequence of the *Actor*, the *Action*, and the *Consequence*.

	The actor	The action	The consequence
Theory	Virtue ethics	Deontology – duty ethics	Utilitarianism
Starting point	Your virtues	Your behavioural norms	Your values

Source: van de Poel and Royakkers (2011): *Ethics, Technology & Engineering.*

Should we (as the actors) primarily be concerned with our own drives and motives, the nature of the actions we take, or the consequences?

The *deontological* approach (from the Classical Greek 'dei' = 'one must') means acting on *moral principles* rather than being guided by *consequences*. Kant believed, and most would agree, that people are moral agents confronted by obligations and duties; in this light he formulated his *Categorical Imperative*: Act only according to that maxim which you can, at the same time, will to become a universal law.

The broad alternative is *consequentialism*, a class of ethical theories holding that the consequences of actions must be central to our moral judgments. *Utilitarianism* is a specific version of consequentialism which suggests that actions are to be judged by the amount of pleasure or pain they cause – 'the greatest happiness for the greatest number' (see Velasquez, Shanks, and Meyer, 1989).

Albert Flores (Flores, 1998) makes the distinction clear: 'Whereas according to the *deontologist* an action is right or wrong insofar as it conforms with our moral duty, according to the *consequentialist* approach the rightness or wrongness of an action is a function of the goodness or badness of its consequences.'

Putting these ideas into relevant contexts requires the citation of particular cases. In recent years there have been many instances where the gulf between two debating positions has been shown to arise from contrasting ethical standpoints; several of the case studies cited in Annex 3 offer instances of the same dilemma. These two brief outlines show the dichotomy clearly:

> A recent case centred on a decision on whether or not to exploit a rich uranium ore deposit in Northern Australia, which was on land held sacred by a small group of aborigines. Different conclusions may be reached when the problem is approached from a consequentialist or a deontological starting point – *what is right* vs. *what is expedient*.

To understand the furore surrounding the disposal of the Brent Spar oil rig, we might agree that the Shell Company took a purely utilitarian point of view in looking at the cost-benefit analyses of the various ways of disposing of an old oil platform. Non-governmental organisations, and Greenpeace in particular, managed to persuade the public that this should be treated as a deontological problem, and to ask: What is the right way, in principle, of disposing of oil platforms?

It should be clear from these examples, and from the many critical debates on ethical theories over hundreds of years, that there is no fixed and agreed definition or interpretation of the realities of the ethical matters that surround us. We cannot escape the responsibility of tough thinking to clarify our minds before engaging in the real world of ethical conflict. *RAE* (2011) is a valuable guide published by the Royal Academy of Engineering to the practical application of ethical thinking.

2.2 APPLYING THE THEORY

To assist in the ethical approach to be taken in future debates, and after discussions with engineers from a number of different engineering institutions and with philosophers specialising in applied ethics, the Royal Academy of Engineering and the UK Engineering Council have together produced a *Statement of Ethical Principles* (RAE, 2014), setting out the values and principles that guide engineering practice and should supplement the codes of practice published by the various engineering institutions. Many of the engineering institutions have followed suit by promulgating for their own members guidance firmly based upon the RAE/EC initiative.

The four principles are:

- Accuracy and rigour.
- Honesty and integrity.
- Respect for life, law and the public good.
- Responsible leadership: listening and informing (RAE, 2014).

and each of these is analysed into several key corollaries:

Accuracy and rigour

Professional Engineers have a duty to ensure that they acquire and use wisely and faithfully the knowledge that is relevant to the engineering skills needed in their work in the service of others. They should:

- always act with care and competence.
- perform services only in areas of current competence.
- keep their knowledge and skills up to date and assist the development of engineering knowledge and skills in others.
- not knowingly mislead or allow others to be misled about engineering matters.
- present and review engineering evidence, theory and interpretation honestly, accurately and without bias.
- identify and evaluate and, where possible, quantify risks.

Honesty and integrity

Professional Engineers should adopt the highest standards of professional conduct, openness, fairness and honesty. They should:

- be alert to the ways in which their work might affect others and duly respect the rights and reputations of other parties.
- avoid deceptive acts, take steps to prevent corrupt practices or professional misconduct, and declare conflicts of interest.
- reject bribery or improper influence.
- act for each employer or client in a reliable and trustworthy manner.

Respect for life, law and the public good

Professional Engineers should give due weight to all relevant law, facts and published guidance, and the wider public interest. They should:

- ensure that all work is lawful and justified.
- minimise and justify any adverse effect on society or on the natural environment for their own and succeeding generations.
- take due account of the limited availability of natural and human resources.
- hold paramount the health and safety of others.
- act honourably, responsibly and lawfully and uphold the reputation, standing and dignity of the profession.

Responsible leadership – listening and informing

Professional Engineers should aspire to high standards of leadership in the exploitation and management of technology. They hold a privileged and trusted position in society – although it is arguable that society does not always behave as though it recognised this status - and are expected to demonstrate that they are seeking to serve wider society and to be sensitive to public concerns. They should:

- be aware of the issues that engineering and technology raise for society, and listen to the aspirations and concerns of others.
- actively promote public awareness and understanding of the impact and benefits of engineering achievements.
- be objective and truthful in any statement made in their professional capacity.

These four principles are, until we internalise them, merely words on the page. In moving from the principles to practice, engineers might well ask how precisely they can be used to guide behaviour? In real-life situations the principles intersect: What should be done when more than one principle applies or when aspects of the situation conflict? Judgement is not a precise process – and there is a danger of oversimplification and of overcomplication. The principles are not self-interpreting rules – reflection is required. They may also imply prohibitions on certain types of behaviour, or may sometimes suggest 'ideals' of behaviour. But in any event they are 'rules' proposed by someone else – to what extent can they be adopted uncritically?

It is worth including here the US National Society of Professional Engineers' 'Code of Ethics', which unsurprisingly covers very similar ground: Engineers, in the fulfilment of their professional duties, shall:

1. Hold paramount the safety, health, and welfare of the public.
2. Perform services only in areas of their competence.
3. Issue public statements only in an objective and truthful manner.
4. Act for each employer or client as faithful agents or trustees.
5. Avoid deceptive acts.
6. Conduct themselves honourably, responsibly, ethically, and lawfully to enhance the honour, reputation, and usefulness of the profession.

Principles, however expressed, will always be applied in specific contexts. The practitioner will always face factors which complicate, such as commercial considerations of cost and contract, personal issues such as the needs and demands of one's colleagues, friends and family, or organisational culture. And of course effective application of the principles requires organisations and individuals to find suitable ways to embed the principles within the culture. There is seldom a clear or black and white choice.

Most engineering decisions can raise ethical questions in your mind. Consider these contexts:

- The selection of raw materials.
- The selection of a supplier or subcontractor.
- The conflicting ways in which the Four Principles may interact.
- The selection of one project amongst many options.
- Personal relationships during the project.
- Pressures of time vs. cost.

All of these situations can give rise – and in practice have given rise – to the need for responsible decisions and may raise ethical quandaries which take the engineer well outside the design manual. What should we do when duties compete? The engineer may find himself conflicted: How is it possible always to act in your employer's best interest if you are not aware of the full complexity of these interests? How can we 'act reasonably' when balancing conflicting interests? And who decides? We sometimes hear that 'safety is always good business' – but not necessarily in the short term; organisations and individuals may gain in the short and medium term by ignoring rare but high-consequence events. Steen and van de Poel (2012) offer a perspective on the demonstration of values through the design process.

The first principle calls for 'accuracy and rigour' but how much accuracy and rigour is it reasonable to apply in any given situation? Accuracy and rigour are 'limitless concepts' – it is open to question how much time, effort, and cost should be expended to achieve higher levels of these indefinable concepts. Engineers are asked to 'identify, evaluate and, where possible, quantify, risk.' This is surely an endless task – and the contract is probably fixed price and time-limited.

Even getting duly paid for one's work may seem to be in conflict with purely ethical perceptions. Work defined and limited by contract may not coincide with what is needed, especially when projects overrun. It is not always easy to define precisely what is required under a contract and to put accurate boundaries around the work. Contracts may thus contain ambiguities, to resolve which an independent assessor may be asked to adjudicate. Contractual arrangements may limit interaction – e.g. in reporting a potential hazard to those affected. (See especially the McDonnel Douglas case in Annex 3.2.)

We are led to the conclusion that careful interpretation of the principles will be required in each context. From lofty principles down to day-to-day decisions and practices – challenges will remain when we balance our duties against the forces upon us. We must use the principles as an initial checklist to identify some key issues, and use brainstorming, be proactive, employ wide discussion, be mindful of emergent issues, and finally, if in doubt or in conflict, ensure at least that your own position is known. Throughout this sometimes complex process every engineer cannot evade the fundamental point that each of us starts from a basis of personal ethical understanding – we understand that moral precepts have significance in our lives and in our thinking, and are therefore relevant to our work as engineers.

Before we leave the discussion of ethical principles we must challenge the usual assumption that responsibilities attach only to individuals. Responsibilities can also attach to those individuals and groups described by Thomas Hobbes in *Leviathan* as 'artificial persons.' By this he meant those who sometimes act or make decisions or issue codes of practice on our behalf, such as lawyers, community groups, professional institutions, church leaders, the military, and so on. Codes of practice to which we are invited to conform should give us pause. Is an action done by someone in the role of a professional engineer in any way different ethically from the same action carried out if the professional role were not part of the picture? If the engineer finds himself guided or hedged by a code of practice issued by the professional body, does this not discourage him from analysing the position as deeply as might have been done if he had not been thus guided? Who then is truly responsible if thereby the overall shape of the profession remains unchallenged by deeper examination? Against a background of professional codes of practice we must nevertheless preserve our individual sense of responsibility and sound judgment; but to do this we must constantly draw on our personal moral experiences. What if the code we sign up to is incomplete or impracticable? Codes are not intended to simplify our lives by enabling us to avoid reflection.

This line of attack reminds us that as engineers we cannot escape the need to engage with the question of the boundaries on responsibility between ourselves and the professional bodies which claim our loyalty; the boundary requires constantly to be challenged if we are to move forward. Wolgast (1992) is an excellent examination of the interface between the individual and the profession.

ETHICAL RESPONSIBILITIES TO CONSUMERS AND CUSTOMERS

In the areas of prescription which concern us here, we have to deal with both legislation and regulation. Legislation sets out what is mandatory in regard to behaviour or performance, and may prescribe sanctions for infringement – e.g. emission standards for road vehicles. Regulations set out the levels of performance regarded as desirable or necessary (and which may be prescribed in legislation) and may have the force of legislation without further discussion. The aspects of product design and development to which legislation or regulation may apply, and which are likely to be most relevant to the work of engineers, will be *product safety, environmental concerns,* and *the management of intellectual property.* We address the question of product safety here and in Sections 4 and 5. Intellectual property management and environmental concerns are covered in Sections 6 and 7.

In designing and developing products our binding obligation to build only *safe* products follows experiences in the real world which have impelled legislators to act. Legislators of course legislate, but always in the light of the last major event; future arisings cannot be addressed ahead of history, nor can we predict and legislate for 'black swan events' (Paté-Cornell, 2012). Safety, quality, and other loosely defined requirements have a cost, a price, and a value or benefit – but who pays? How much is it worth paying? Who decides? How can we resolve conflicts? The worldwide position on regulation is complex – as an engineer, how much can you be aware of? How do you choose your priorities? These questions are real, and the answers may or may not be obvious, but in the context of a fixed price contract it is clear that the engineer has to produce answers which reflect, in effect, the best that can be done within the resources available.

These challenges for the engineer are as old as the hills but have been made more complex by the proliferation of legislation and regulation in recent decades, following incidents whose publicity

DOI: 10.1201/9781003208433-3

concerned consumers. In general, legislation increases costs while asserting, without proving, supposed benefits; if assertions are made claiming a benefit (such as a fall in deaths on the road following the introduction of mandatory seat belts) the cost is borne by the manufacturer and passed on to the customer; negative impacts – such as the tendency of drivers to drive faster when seat belt was introduced – are ignored. Nor is the value of a life saved brought into the calculation.

It is the operation of a healthy competitive marketplace which truly delivers the goods, and in such a market the relationship between the producer and the consumer is critical. Consumers often react assertively when making their choices – not many are ignorant or neutral or apathetic; designers must respond accordingly. The designer's task is already made complex by the many standards affecting product performance. The overall task will address at least: design for manufacture, design for form, fit and function, design for environment, design for safety, design for sustainability, design for electromagnetic compatibility (EMC), design for efficiency, design for disposability, and perhaps other requirements. How are these to be incorporated in a design to be delivered within a fixed time and budget? One answer, following the experience of many decades of constructive competition, is that progress in the applicable technologies and the operation of the market ensures that products do indeed improve consistently from one generation to the next – provided that the relationship between the producer and the consumer operate on an agreed ethical basis.

Three different theories of the ethical duties of manufacturers in regard to product safety have emerged in recent times:

- The 'contract' view
- The 'due care' view
- The 'social costs' view

The 'contract view' of a supplier's duty to customers proposes that the relationship is purely a contractual one, in other words the consequence of a sales contract freely entered into by both parties which makes its stipulations clear and must by law be adhered to. This idea conforms broadly to Kant's proposition for a deontological basis for ethical thinking; we would want to argue that people and organisations have a duty to do what they have contractually agreed to do, because failure to do so would prevent the use of contracts becoming universally accepted.

The contract view of the relationship can be criticised on three main grounds: firstly, in practice the manufacturer and the consumer cannot meet on common ground because the former has inevitably more detailed knowledge of the product than any consumer can practicably obtain; secondly, manufacturers do not usually deal with customers face to face but through middlemen, dealers, or retailers, thus weakening any bond definable in a contract; and thirdly, disclaimers of various kinds are often in place, reducing the leverage of the consumer.

The 'due care' view attempts to address these difficulties by acknowledging that manufacturers and consumers cannot meet as equals, and that therefore the manufacturer's duty must embrace a special concern for customers' well-being, beyond the scope of a contract. This duty of care should also extend beyond the immediate customer to others who might be injured by the product. But this view cannot take into account the costs of such extended care or the interlocked nature of the decisions on cost and risk which the producer would be required to take. Who decides?

The 'social costs' view takes a still broader perspective and extends the duties of producers to cover the costs of any injuries caused by any defects in the product, even where all due care was taken during its manufacture. Such strict liability is based upon utilitarian principles and holds that producers must bear the costs of all outcomes regardless of the precise locus of identifiable fault. Manufacturers might regard this as unjust, given that they could not be expected to foresee all and any of the circumstances in which the product might be used; in this context is it reasonable to expect them to be ready to meet all such costs and consequences?

These three views are explored in more detail in (Velasquez, 2012, section 6.2) and (Hasnas, 2008) and in several of the case studies in Annex 3. Here we may note that it will never be possible, and it would be undesirable, for one view to obtain universally; the context and prior law will determine how much will depend upon earlier experiences in this crucial producer–consumer relationship and upon how well the market operates. In view of the problems in defining safety and interpreting those definitions in multiple real-life contexts, neither producers nor consumers should expect clean-cut positions. We note a swing from the old principle *Caveat Emptor* (let the buyer beware) towards *Caveat Vendor* (implying that the producer must be ready to carry the can).

For the engineer the challenges will include:

* What are the relevant requirements in the *design*, *develop*, and *build* processes?
* How flexible or adaptable are they? How does one find out what they are?
* Can they be challenged or modified if necessary? If so, who carries the responsibility?
* How can one incorporate the requirements into the design process?
* 'Where does personal responsibility begin and end?'

The engineer must therefore:

* Read and understand the contract;
* Read and understand the required standards;
* Define and understand the terminology;
* Seek clarity by questioning;
* Ensure that guides and limits are set and are clear;
* Decide how he will test designs against the specification; and
* Ensure he understands the interfaces between his responsibilities and those of others – the customers, the company, the team.

A report from the Hastings Institute by Powers and Vogel (Vogel, 1980) presents a useful study with recommendations on the structure and content of ethics education for business managers.

To illustrate the notion that it is questionable whether an ethical perspective on the supplier's duties can form a comfortable basis in analysing behaviour, the present author can outline a case in which he was personally involved, and was indeed in part culpable.

In the context of a commitment by the employer to deliver certain stipulated items of equipment on time and cost, it became clear at a late stage, first, that the company had misled the client – and perhaps also itself – as to its ability to meet the contractual dates; second, that this information had been withheld from the client until a very late stage; and third, that we had not put into the project the resources that, on any reasonable forecast, were essential to in-house support of the project.

It fell to me, as the supplier's Project Manager, to chair the uncomfortable meeting at which it became clear to the client that we were going to let him down. The emotional temperature of that meeting was unforgettable. The subsequent renegotiations cleared the air to some extent. I was reproved by my employer for having revealed the internal weakness that the funding for the project had not been maintained. The lessons I learned then, about open dealings with clients and clarity over spend against budget, are with me permanently, but it is fair to say that the ethical issues involved were not scrutinised by me or by the company at the time. If any degree of ethical awareness had been in the forefront of our minds, an uncomfortable and unanswerable question now poses itself: Would the sequence of events have been any different? Perhaps; but to my recollection our only concerns at the time related to commercial pressures and personal culpability.

After a 40-year lapse of time one can only reflect on the whole issue in hindsight and, in so doing, hope to become wise. Velasquez (1983) provides an illuminating attack on the notion that businesses can be found 'responsible' in the true sense of that word, and challenges us to attribute responsibility, together with blame, where it truly should be placed – on the shoulders of individuals with moral principles.

SAFETY AND RISK

What are the bounds on product safety? This is an issue of great concern to regulators and legislators alike, because many producers make and sell products which might be considered 'unsafe' under some conditions. All products are designed, built, and marketed in competitive conditions. How far must manufacturers go to make their products 'safe'? Companies know more about their products and processes than their customers. Do they have a duty to protect their customers from *all and any possible harm*? What about injuries which no one could have foreseen or prevented? Product safety is perhaps the most basic but also most contentious issue which arises in the debatable land between producers and consumers. It is worth stating right at the start that safety always has a *cost*, a *price*, and a *value*; the cost is determined by the *producer*, the price by the *market*, and the value by the *customer*.

This being so, for how long should any manufacturing company's liability to the customer or consumer last? If as a design engineer you are to be responsible for the safety of any products during your career, you would be wise to ensure that your responsibilities are defined and bounded. Are tobacco companies to be made liable *retrospectively* for the 'safety' of their product? What is 'safety' in this context? What does 'safe' mean? If safety has a price, a cost, and a value, how are these to be assessed and agreed? If the context constrains the design, to what extent? And who decides?

Safety is *the state of being 'safe'*, the condition of being protected against physical, social, spiritual, financial, political, emotional, occupational, psychological, educational, or other types of consequences of failure, damage, error, accidents, harm, or any other event which could be considered undesirable.

Safety can also be defined to be 'the control of recognized hazards to achieve an acceptable level of risk.' This may imply being

DOI: 10.1201/9781003208433-4

protected from certain events or from exposure to something that may cause health or economic losses. See Manuele (2010) on the concept of acceptable risk.

The vagueness and generality of the definition given above, which may be comforting but is of little value in any practicable sense, reminds us that the word 'safe' is used so loosely in so many diverse contexts as to be of no value in the design of engineering systems. A condition of 'zero risk' is unattainable in the real world, and could not be asserted with any convincing evidence. To enable us to address acceptable levels of risk in technologically complex contexts an *upper risk limit* may be specified, beyond which operation is unacceptably dangerous. A *lower limit* may also be identified below which the cost of implementation may not be commercially viable. In between the two levels is a band of risk which must be justified by argument and demonstration; the identified risks in operation must be reduced to a level which is *ALARP – as low as reasonably practicable*. It is worth remarking also that safety features may reside in the *processes* by which the product is manufactured as well as in the final *product*. A *safety case* (see below) may be required to enhance confidence levels.

The objective of *functional safety* is freedom from unacceptable risk of physical injury or of damage to the health of people either directly or indirectly (through damage to property or to the environment). *Functional Safety Standards* govern the safety-certifiable elements of systems or equipment upon which *correct operation* depends – and include the safe management of operator errors, hardware failures, and environmental changes.

If a system is to be declared 'safe', it must be feasible to define what is meant and how the assertion is to be proved or demonstrated. We must therefore define safety, be able to translate the definition into *standards*, and then show through a safety case that we conform to those standards. In the context of the four RAE/EC ethical propositions (Accuracy and rigour; Honesty and integrity; Respect for life, law, and the public good; and Responsible leadership) we will want to assert, with integrity and with full responsibility, that an appropriate safety level has been reached. But as we have said, in real-life situations the principles intersect or interact; what is to be done when more than one principle apply, or when they conflict? It is clear that this can never be a precise process; there is the danger of oversimplification; the principles are not self-interpreting but must be closely examined before they can be applied. Some behaviours may be prohibited; some others are simply ideals. The public often reacts unfavourably when it is pointed out that safety has a price, which is paid by the consumer; see, for example, the Ford Pinto case at Annex 3.6. There will always be a conflict between our demand for 'safe' products and our wish for lower costs. The case of the Golden Gate Bridge (Figure 4.1) safety barriers demonstrates this conflict dramatically (see Annex 3.7 for the complex arguments for and against the installation of safety barriers).

Clarity in these matters is especially important in regard to *safety-critical systems*, whose failure could result in loss of life, significant property damage, or damage to the environment. Examples are air traffic control, aircraft control systems, space vehicles, medical equipment, modern information systems, nuclear systems, banking systems, financial systems, and many more. A modem *heart pacemaker* is a computer with specialized peripherals; *fighter aircraft* rely heavily on computer networks, as do modern cars; and many *defence facilities* are actually distributed computer systems. Our concern both intuitively and formally is with *the consequences of failure*. If the failure of a system could lead to consequences that are determined to be unacceptable, then the system is *safety-critical* (Addagarrala and Kinnicutt, 2018). In essence, a system is safety-critical when we depend on it for our well-being. Hardware *and* software may be implicated. Consider software compilers: A suite of software may of

Figure 4.1 The Golden Gate Bridge.

itself satisfy safety-critical criteria – but in order to run it has to be compiled. Two instances may be cited to illustrate the significance of such systems:

A problem arose with the primary protection system for the Sizewell B nuclear power reactor. This system was implemented in software and was designed to achieve a reliability requiring fewer than 10^{-4} failures per demand. Once the system design was complete, it used more than 650 microprocessors and 1,200 circuit boards, and the software was over 70,000 lines long. When system tests were carried out, the system passed only 48% of the test cases. The system was reported to have failed the other 52% but, in fact the real problem was that, for many of the tests, it was not possible to determine whether the test had been passed.

On 26 October 1992, the ambulance service for the city of London switched from a manual dispatch system to a computer-aided dispatch system. The changeover was made all at once so that the computerized system was expected to operate for the entire coverage area. The system worked initially, but a complex sequence of events led to the system being essentially non-operational as the demand increased during the day. Since ambulance dispatch was severely delayed in many cases, there is good reason to think that deaths or injury resulted from the failure.

We focus on safety-critical systems because they are becoming more ubiquitous, they often incorporate software as well as hardware, they are becoming more complex, and they are becoming more troublesome. Obvious systems in everyday life are the flight controls of aircraft, the braking system on the car, nuclear reactor controls, air traffic controls, railway signalling systems. And there are many less obvious examples which have the potential for very high consequences of failure and should probably be considered safety-critical – the 999 telephone system, banking and financial systems, the water and the electricity supply system. These are all computer controlled and monitored, and if they failed – or, worse, were compromised or brought down by hostile activity – would remind us of our dependence on good system design. We are reminded of the importance of systems thinking at the design stage. Ethical issues (involving perhaps safety criticality, environmental concerns, intellectual property issues, conflicts in project management, etc.) might be buried anywhere within a complex system. It is not surprising that tracking down the abuse in the Volkswagen environmental software (see Annex 3, case study 1) proved so difficult.

Bowen (2000) proposes a code of practice for designers of safety-critical systems. He suggests that engineers working on safety-related systems should:

- Take all reasonable care to ensure their work and the consequences of their work cause no unacceptable risk to safety;
- Not make claims for their work that are untrue, or misleading, or are not supported by a line of reasoning that is recognized in the particular field of application;
- Accept personal responsibility for all work done by them or under their supervision or direction;
- Take all reasonable steps to maintain and develop their competence by attention to new developments in science and engineering relevant to their field of activity, and encourage others working under their supervision to do the same;
- Declare their own limitations if they do not believe themselves to be competent to undertake certain tasks, and declare such limitations should they become apparent after a task has begun;
- Take all reasonable steps to make their own managers, and those to whom they have a duty of care, aware of risks they identify; and make anyone overruling or neglecting their professional advice formally aware of the consequent risks; and
- Take all reasonable steps to ensure that those working under their supervision or direction are competent; that they are made aware of their own responsibilities; and they accept personal responsibility for work delegated to them.

Many engineers are involved in design projects with embedded software and therefore must consider the contributions of hardware *and* software to performance and product safety margins, especially in safety-critical systems. The analysis then becomes all the more important in view of the nature of coding practices. Furthermore, Bowen (1996) places the discussion in the context of formal software methods and standards. Standards for safety-critical software are registered on a scale of five discrete levels of safety integrity. An integrity level of 4 is 'very high'; the term 'safety-related' is used to refer to integrity levels 1 to 4; a system at level 0 is not safety-related. A deeper analysis will assign an integrity level to each component of a system, including the software.

Our personal responsibilities, therefore, require us to ask these questions in regard to every system we develop: Might it present hazards to safety? How do we identify the possible hazards? What can we do to reduce hazards to an acceptable level? How should we define safety and then assert that the developed system is safe?

To enable us to make justified assertions about the safety of our designs we need confidence in the regulators, approved procedures in design and operations, management systems which inspire confidence, proper training, and, perhaps above all, a sense of personal responsibility. To achieve safety as properly defined in design, manufacture, and operation, everyone must recognise his/her own contribution to the maintenance of standards – and, importantly, a commitment to the raising of those standards – in other words, a safety culture throughout the organisation.

If the system might impact on the safety level we must conduct a detailed risk analysis, asking: How likely is it that an error in the system will result in a particular hazard? How likely is that hazard to actually cause an accident? What would be the consequence of an accident? It is also essential that 'ownership' of each risk be identified.

The development of a safety-critical system should aim to minimise the loss of human life or serious injury by reducing the risks involved *to an acceptable level*. This is normally considered an overriding factor; the system should always aim to be safe, even if this adversely affects its availability. It is the responsibility of the engineering team and the management of the organisation, firstly to define 'acceptable', and then to ensure that suitable mechanisms are in place and in use to achieve this goal for the lifetime of the product. See Addagaralla and Kinnicutt (2018) for a good recent analysis of the standards and different ground rules to be followed in critical software development practices in different industries; see also IEEE Computer Society (1999) and Figure 4.2. Another important source of definitions and standards is the International Electrotechnical Commission's publication IEC 61508, *Functional Safety of Electrical/Electronic/Programmable Electronic Safety-related Systems*; these are guidelines on how to apply, design, deploy, and maintain automatic protection in systems which will be safety-related.

It can be argued that, in this matter as in all matters, truth can be attained not only through scientific knowledge, but also technical skill, prudence, intelligence, and, supremely, wisdom born of experience. A balanced team will try to see the design from all of these perspectives, and this will imply careful selection of team members and exposure of the design across team boundaries. Only thus can agreement be reached on the appropriate definition of 'acceptable.'

Standard	Description
Quality Systems - Model for Quality Assurance in Design/Development, Production, Installation and Servicing. ISO9001/EN29001/BS5750 part 1	This is the recommended minimum standard of quality system for software with a safety integrity level of 0, and an essential prerequisite for higher integrity levels.
Functional Safety : Safety Related Systems. IEC1508	A general standard, which sets the scene for most other safety related software standards.
Railway Applications: Software for Railway Control & Protection Systems. EN50128	A standard for the railway industry.
Software for Computers in the Safety Systems of Nuclear Powers Stations. IEC880	A standard for the nuclear industry.
Software Considerations in Airborne Systems and Equipment Certification. RTCA/DO178B	A standard for avionics and airborne systems.
MISRA Development Guidelines for Vehicle-based Software	Issued by the Motor Industry Software Reliability Association for automotive software.
Safety Management Considerations for Defence Systems Containing Programmable Electronics. Defence Standard 00-56	A standard for the defence industry.
The Procurement of Safety Critical Software in Defence Equipment. Defence Standard 00-55	Detailed software standard for safety critical defence equipment.

Figure 4.2 Some published standards for safety-critical software.

Sometimes this can lead to startling conclusions: In the Ford Pinto case (see Annex 3.5) the company argued that the cost of redesign to remove a questionable feature would exceed the total likely payouts in the event of the deaths of users of the unimproved model and should not therefore be undertaken. But decisions of this broad kind (in effect: 'How much is a human life worth?') are taken all the time in many contexts without disturbing our perceptions when we make choices as consumers.

In debating the acceptable level of risk in a design we must bear in mind that risk is a function of the frequency (or likelihood) of a hazardous event and the consequent severity of the event. Zero risk can never be reached; safety must be considered from the beginning of the design task, and non-tolerable risks must be reduced so that they are *ALARP*.

The safety case methodology provides a systematic and complete way to show compliance to one or more standards. The methodology was established in industries which deal with functional safety of computerized automation in nuclear and avionics. A safety case is a structured argument presenting evidence intended to demonstrate that a system is as safe as reasonably practicable; it should satisfy the authorities that specific safety claims can be met and that risks are ALARP.

In defence matters, Def-Stan 00-56 Issue 4 addresses 'Safety Management Requirements for Defence Systems', including the creation of safety cases. The definition in the Standard states: A safety case is a structured argument, supported by a body of evidence, that provides a compelling, comprehensible, and valid case that a system is safe for a given application in a given environment. Bloomfield and Bishop (2010) offer a clear guide to modern and thorough safety case creation.

UKSPEC 2020), published by the Engineering Council UK, lists six principles to guide and motivate professional engineers and technicians in identifying, assessing, managing, and communicating about risk. For more information, see www.engc.org.uk/risk and UKSPEC page 34. The six principles are:

- Apply professional and responsible judgement and take a leadership role.
- Adopt a systematic and holistic approach to risk identification, assessment, and management.
- Comply with legislation and codes, but be prepared to seek further improvements.
- Ensure good communication with the others involved.
- Ensure that lasting systems for oversight and scrutiny are in place.
- Contribute to public awareness of risk.

The case of the Boeing 787 Dreamliner (Figure 4.3) shows how difficult it may be to identify and isolate areas of risk in a design. In January 2013, lithium-ion batteries on a Dreamliner caught fire shortly after landing at Boston airport following a flight from Tokyo. Two other B787s experienced similar problems in the same week – all owned by Japanese airlines. The aircraft was three years behind schedule when entering service. This had been deemed by the FAA

Figure 4.3 Boeing 787 Dreamliner. *By pjs2005 from Hampshire, UK - Boeing 787 N1015B ANA Airlines, CC BY-SA 2.0,* https://commons. wikimedia.org/w/index.php?curid=71147495.

a 'safe' aircraft – but the whole fleet was grounded. Federal Aviation Authority regulators approved flight tests of Boeing's 787s while the company and government investigators tried to pinpoint the cause of the burning batteries. Incidents continued to arise but the underlying cause was not understood. The authorities were pessimistic about how quickly the planes might resume commercial flights.

Doubts had earlier been expressed in many contexts about the safety of Li-ion batteries in high-power applications, and the risks had previously been recognised. In 2006 fire destroyed an aerospace building in Arizona after laboratory tests on a Li-ion battery got 'out of control.' In 2010 a Boeing 747 crashed in Dubai when Li-ion batteries in the hold caught fire. The potential dangers of Li-ion batteries in power applications had therefore been known for some time. Rumours spread; a documentary TV programme cited engineers on the Dreamliner project expressing serious concerns about its safety, although it is not known whether these doubts related to the batteries. Engineers were also filmed saying they feared colleagues working on the aeroplane were using drugs. Such revelations threaten to reopen concerns about the Dreamliner. The allegations were disputed by Boeing and the FAA maintained that: 'This is a safe aircraft.' The story shows how easy it is for doubts to arise, fanned by the flames of gossip and innuendo; the importance of a proper risk analysis shared by all parties is emphasised.

THE DEBATE ON AI AND AUTONOMOUS SYSTEMS

The continuing development of artificial intelligence (AI), autonomous systems (i.e. systems operating without human intervention), and machine learning has inevitably led to a welcome debate on the basis of our approach to thinking about ethics in this context. If society is expected gradually to become accustomed to the use of these systems, appropriate decision-making will have to be developed during the design, development, and operation of such systems; knowledge and experience must be widely shared so that lessons can be learned; and ethical problems must be addressed.

A starting point for this discussion might be the 'trolley problem', a well-known thought experiment in ethics. A runaway trolley is accelerating down the railway tracks. Ahead, on the tracks, there are five people tied up and unable to move. The trolley is headed straight for them. You are standing some distance off in the train yard, next to a lever. If you pull this lever, the trolley will switch to a different set of tracks. However, you notice that there is one person tied up on the side track. Your options are: Do nothing, and the trolley kills the five people on the main track; or pull the lever, diverting the trolley onto the side track where it will kill one person. What do you choose? A different version of the same challenge has the observer standing on a bridge overlooking the scene; by pushing a nearby stranger over the edge and onto the track both sets of prisoners can be saved; but this version presents a much greater degree of moral challenge for the observer.

For many people the sheer unreality of this problem understandably makes it easy to back away without a decision. The problem arises, however, in a starker form for a team of engineers collaborating in designing the AI system to provide directions for a driverless vehicle. Such vehicles already operate on our roads, but even educated people understand very little about the underlying logic

DOI: 10.1201/9781003208433-5

structures and processes which have guided the developers; and more potently, about the confidence we should need to give us courage to stand in front of such a vehicle heading towards us.

As robots become more autonomous (if indeed they do so) society must develop rules to manage them. In the film *2001*, the ship's computer, HAL, faces a dilemma. His instructions require him *both* to fulfil the ship's mission (investigating an artefact near Jupiter) *and* to keep the mission's true purpose secret from the ship's crew. To resolve the contradiction he tries to kill the crew.

Autonomous machines, as they become more 'intelligent' and more widespread, will be in the position of making life-or-death decisions in unpredictable situations, thus assuming, or at least appearing to assume, moral agency. Autopilots have made planes safer, but the captain and crew always carry ultimate responsibility. Should we insist that the driver of a road vehicle be present at all times to take back control in an emergency? Autonomous vehicles may be safer than vehicles with a human controller in *most* circumstances, but many road users will need to be persuaded that this can be true in *all* circumstances.

Robotic surgery has moved into closer focus in recent years. In theory a robotic surgeon is well suited to the most delicate surgical procedures; it can enter body parts human fingers cannot reach and can more delicately cut without the risk of hand tremors. Robots can already perform intricate keyhole surgery. Stitching one part to another, being repetitive and precise, has the potential to be automated. Robots so far have been controlled by surgeons sitting at a console. These machines lack machine learning and true AI. But in future…? A surgeon was quoted recently as saying: 'To tell a patient that I will press a button and everything will be done by a robot while I have coffee may not be acceptable to many.'

Isaac Asimov produced in 1942 'three laws of robotics', requiring robots to *protect humans*, *obey orders*, and *preserve themselves*, in that order. What happens, or should happen, when robots are put into situations which call for ethical decisions? Are Asimov's criteria adequate? How can we be sure that software intended to deliver such criteria did in fact do so? How could a robot or a driverless car be programmed to make the 'right' choice when either option presents ethical dilemmas, involving, e.g., human lives? Where might the costs fall if a driverless car damages people or property as a result of choices it has been programmed to make? Where does the buck stop – with the designer, the programmer, the manufacturer, or the operator?

Is it possible to programme an AI machine to avoid or escape from any situation which has occurred suddenly and for which it had not been programmed to learn an appropriate response, or had not experienced often enough to learn a response for itself?

One cannot avoid asking these almost unceasing questions and hoping that in time, with experience and growing wisdom, answers will emerge. It seems that we shall need autonomous vehicles with an ethical compass; but that would have to be learned. If an engineer signs up to (for example) the IET's Code of Practice, can he be expected to sign on behalf of any autonomous machine he might ever have to design and build? Is he *in loco parentis* – as with his own children – until the robot is old enough to sign for itself?

On the battlefield AI robot soldiers would not commit rape, burn down a village in anger, or become erratic decision-makers amid the stress of combat. It must be assumed, however, that sooner or later autonomous weapons will be developed which have the ability to make their own decisions in the battlefield. That little word 'decision' should make us think. If an outcome follows an autonomous analysis of ever longer data strings, to what extent can it be called an independent decision? There will always have been a precursor situation, and earlier a whole string of such situations, leading back to a first-generation AI programme. Where was that leading decision – the first choice between an X and a Y, or between a 1 and a 0? *Roborace* is a fledgling motorsports series using autonomous race cars. Its beta-test season was being carried out in October 2020; one of the cars 'decided' to drive into a wall from a standstill. No one was injured – but this outcome was certainly neither intended nor expected.

We must be aware of the ethical challenges in this otherwise unsupervised sequence. We may be content to know that AI can be, and is being, used in surveillance, information gathering, and analysis; but it might sooner or later be used in live weapons systems. If AI systems themselves can have no ethical compass, no means of asserting that X is wrong, and could not even assert that an alternative Y would be right (even if they might avoid an unwise choice), the true basis of the ethical underpinning must surely go back to the designer.

AI learning depends for its development on the accumulation of very large quantities of data, on the basis of which sound conclusions can supposedly be drawn, perhaps with growing confidence as each new conclusion succeeds its predecessor. At what point can the reliability of the data on which each decision has been built be challenged? We may have to generate new definitions of words like 'intelligent' and 'decision' for use in these debates. And we may never agree on the risk and safety elements. The safety boundaries seem to become uncertain; where is an ALARP risk level in all this?

On a lighter note, a news item from 2005 reported that a South Korean woman fell asleep on the floor and was 'attacked' by her robotic vacuum cleaner. The woman had to call firemen after the cleaner sucked up her hair and she was unable to extricate herself. It required two paramedics to free her from the machine's nozzle. Beware unintelligent robots!

Bryson (2018) offers an excellent guide through this minefield. But there are many questions which currently have no agreed answers; in fact, all we have are the questions. In the current competition between the USA and China for global dominance, it is easy to convince oneself that we are already engaged in an AI arms race. It remains to be seen whether the huge issues facing us will be satisfactorily resolved by future debate, designs, and practices. See also BSI (2016) and RAE (2009).

ETHICS IN THE MANAGEMENT OF INTELLECTUAL PROPERTY

Intellectual property (IP) in its broadest sense may be defined as *any asset under identifiable ownership which cannot be immediately valued in financial terms (as can, for example, assets on the balance sheet) but must be protected and managed carefully to generate value and revenues over time.* See Annex 4 for a brief outline of the categories of intellectual property. This section is concerned with the potential ethical pitfalls which may be encountered in the management of IP.

Intellectual assets tend not to have a calculable financial value, as their long-term value will depend upon how well the available market is created and sustained against competition. If carefully exploited they will have a substantial impact on future cash flows and profits, but their aggregate value over time cannot readily be assessed. Therefore, business leaders will be aware of the need to develop and market their intellectual assets, think strategically, and engage in activities concerned with *knowledge management* in the broadest sense. Indeed there is an ethical duty on executives to manage company resources, including the knowledge base, for optimum outcomes; to fail to do so will imply a failure to maximise the return on assets, and perhaps failure to reflect the innovative and creative skills of the workforce.

The originator or owner may decide not to seek protection for IP. The costs of comprehensive cover may be prohibitive, especially for small businesses. It may make sound business sense to exploit new innovation rapidly to exploit a market opportunity, relying upon fast action and continual inventiveness to keep the business ahead of the competition. These may be respectable reasons for deciding not to patent, and they remind us of the need to manage IP in the context of a knowledge management plan, rather than simply *ad hoc.*

DOI: 10.1201/9781003208433-6

Usually, however, a patentee intends to exploit his own IP during the period allowed by patent; companies will aim through ownership of the intellectual property rights (IPR) to earn revenue from their own knowledge base. In some circumstances, however, instead of embarking upon expensive research and development it may be sensible to licence-in intellectual property developed elsewhere. Yet another source of revenue may arise if a licence can be negotiated with another organisation, which is thereby free to use the IPR to manufacture under licence for a stipulated period of time, and/or address a portion of the market.

Ethical issues in IPR management may arise in several aspects of the management task, for example, regarding *ownership, copying, distortion, abuse, assessment of value*, to name only a few. An engineer may knowingly or unknowingly incorporate into his work elements which were created and protected elsewhere; conversely, elements of his work may be used without acknowledgement by others. There have been cases where companies have deliberately used the IP of others without acknowledgement and therefore without a fee. All these aspects and more keep patent lawyers happy.

When in 2002, for example, Dyson sued GE in the USA for copying the cyclone mechanism, GE paid £4 million in damages. The settlement, believed to be higher than any court award in a UK patent case, ended a long court battle following the launch of Hoover's triple vortex cleaner in early 1999. Dyson said that it offered to settle its claim for £1 million in 1999 but Hoover had refused; Dyson then went to the High Court, accusing its rival of infringing its patent. The court ruled in Dyson's favour in October 2000 and ordered Hoover to stop selling the triple vortex immediately. The CEO of GE was reported as saying that, with hindsight, GE's strategy should have been to become sole licensee in the relevant US territories and simply to have kept quiet while marketing their own version.

For many decades it was believed that Japan flagrantly copied Western products in order to help her rebuild her industries after WWII; in her defence it was often stated that culturally the peoples of Eastern nations lacked the innate ability to innovate as the West, led by Britain, had done since the Industrial Revolution. The improvement in technological capability in China and Japan in recent decades shows that this argument was false.

Recently the contrary view has been expressed that for too long the IPR of the advanced West has in effect been 'food for rich countries and poison for poor countries.' On one side, in the developed world, there exists a powerful lobby of those who believe that protected IP is good for business, benefits the public at large, and acts as catalysts for technical progress; and that only by exercising the right to restrict access can be expensive research and development programmes (e.g. to develop powerful new medicines) be afforded. On the other side, in the developing world, there exists a vociferous lobby of those who believe that current IP practice cripples the development of local industry and technology, harms the local population, and benefits only the developed world. They argue that intellectual property – patents and copyrights – have become controversial. Teenagers are sued for pirating music; AIDS patients in Africa are dying due to lack of ability to pay for drugs that are priced high to satisfy patent holders.

Engineers should form their own views on this important issue. Are patents and copyrights essential to thriving creativity and innovation? Do we need them so that we may enjoy fine music and good health? Does this set of priorities, on the other hand, lead us inevitably to compromise the access to new medicines and techniques amongst poorer nations? So-called intellectual property can thereby be seen as an intellectual monopoly that hinders rather than helps the competitive free market regime that has delivered wealth and innovation to our doorsteps. Boldrin and Levine (Boldrin, 2013) have sound coverage of both copyrights and patents and their impact worldwide; the authors conclude that the appropriate policy to be followed should be to eliminate patents and copyright systems as they currently exist.

In the UK, the Department for International Development published in 2002 (UK DID, 2002) a survey of current views on the impact of IP management upon aspects of international development; the debate continues, not very vigorously. Executives and governments must decide on the appropriate strategy to be adopted with regard to IPR; their decisions inevitably have strong ethical undertones, but doing nothing is not an option. Politically this is not often regarded as a matter to be accorded high priority, nor is public opinion vocal, but the ethical aspects are real and must not be ignored; engineers who care about the impact of their work on the wider well-being of humankind (see Section 7) cannot avoid forming a view and perhaps taking sides.

As recently as early 2021, during discussions on the distribution of COVID-19 vaccines, the ruling council of the World Health Organisation debated a motion to 'force' the pharmaceutical companies which had developed the new vaccines (very successfully and rapidly) to waive their patents so that COVID vaccines could be developed and distributed more widely at cost. But even the WHO recognises that patent protection is crucial in the development of effective medicines; drug companies must be allowed to offset the high cost of research and development, given that many proposed drugs never complete the course through to widespread use. In the case of Covid, moreover, the pharmaceutical companies have played major roles in the international effort to get vaccines delivered quickly to poorer countries. So one must be cautious before assuming that patent protection leads to dangerous monopolies.

ENVIRONMENTAL MATTERS AND SUSTAINABILITY

A developing awareness of wider ethical perspectives in all aspects of engineering, from design through to delivery and use, must lead immediately to the question of our obligations to the environment, the natural world, and future generations. What, ultimately, is the source of these obligations? What is to be done when the rules of conduct do not explicitly tell us how to act?

It is generally accepted that environmental concerns over the impact of technology must be incorporated into the design process, given the adverse impact of the extraction processes of some raw materials and uncertainty about the long-term effects of some technologies on the environment. Engineers and others should accept the responsibility to care for the environment and expand the range of questions they ask themselves when making decisions. In light of this, decisions need to be made on a wider basis than hitherto. Awareness of the relevant statutory requirements also becomes an essential feature of the knowledge base.

The National Environmental Policy Act (NEPA) 1969 in the USA was among the first laws to establish a broad national framework for protecting our environment. NEPA's basic policy is to ensure that all branches of the US government give proper consideration to the environment prior to undertaking any major federal action that significantly affects the environment. NEPA requirements are invoked when airports, buildings, military complexes, highways, parkland purchases, and other federal activities are proposed. Environmental Assessments (EAs) and Environmental Impact Statements (EISs), which are assessments of the likelihood of impacts from alternative courses of action, are required from all federal agencies and are the most visible NEPA requirements.

DOI: 10.1201/9781003208433-7

Concern for the environment leads directly to a concern for the *sustainability* of our initiatives. The World Commission for Environment and Development (WCED), in the Brundtland Report, defines *sustainable development* thus:

> Development that meets the needs of the present without compromising the ability of future generations to meet their own needs.

(Kono, 1987)

Another definition:

> Sustainability is the capacity to continue to function into the future; essentially a new way of seeing the world based on justice and shared responsibility.

This is perhaps too vague to give us a handle on what we should actually do now; there will always be valid questions about the extent of our practicable reach for sustainability. Always we come up against project pressures in fixed price contracts, and in the operation of competitive markets, which might seem to give sustainability a low priority amongst other factors in decision-making. The allocation of today's wealth between now and the future is a national matter, to be decided by our leaders in the political arena. All we can do as thoughtful engineers is to ensure as best we can that ethical matters are given priority. The question is inevitably raised: Who ultimately is to be responsible for the decisions made, given that few people will not be aware of the need for a sound balance? The buck stops way over my head – but perhaps it starts with me (Figure 7.1).

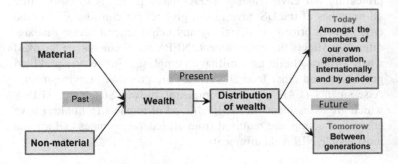

Figure 7.1 Model of the sustainability goal.

We should also consider the application of sustainability in the context of investment decisions. When it comes to choosing good investments, the guiding principle is profit: Will a new capital investment generate a revenue stream greater than the cost of acquiring it? If yes, the jobs that accompany the capital investment will be sustainable, at least as far as we can reasonably surmise. If no, the project will lose money and will either end quickly or will require subsidies funded by adding costs elsewhere in the economy. Again we are here seeking to balance the present against the future.

'Sustainability' does not mean using fewer resources or cutting energy consumption, though it may involve those things. It means value creation in a competitive marketplace where the concept of value can be expanded to include (but not consist solely of) natural and environmental capital. A profitable investment is one where, after all costs are paid (including environmental costs), the outputs are worth more to society than the inputs, including the labour costs. Profitable investments are sustainable, unprofitable investments are not. Friedman (1970) would recognise this perspective.

A starting point might be to consider to what extent we might build environmental and sustainability concerns into the design process. Our first attempts will rightly be cautious, but with growing confidence we might become more confident to reach for more ambitious goals without jeopardising project demands. Experience suggests that competitive contracts operating to deliver products into competitive markets can lead, and in the past has led, to dramatic changes in effective designs which meet challenging targets. Legislation may be needed to set the levels of various environmental targets; engineers can then be left to address the realities and meet market needs. See van de Poel and Royakkers (2011) section 10 for a fuller discussion of some sustainability issues (Figure 7.2).

How sustainable is the world economy, given the inexorable rise in population? Evidence from recent generations suggests that we can be optimistic about our ability to solve current problems so as to meet current and future needs sensitively, provided that wise leadership can deliver the economic growth needed to maintain our efforts on sustainability and environmental care. Whether it can continue to do so will depend upon the identification and selection of leaders of the calibre needed to address the challenges consistently, and the emergence of new technologies to assist and direct our choices.

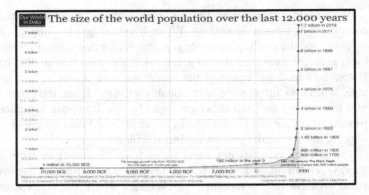

Figure 7.2   World population. CC BY See http://themasites.pbl.nl/ tridion/en/themasites/hyde/basicdrivingfactors/population/index-2. html

Section E3 of UKSPEC requires candidates to be able to show that they undertake engineering activities in a way that contributes to sustainable development; this could include an ability to:

- Operate and act responsibly, taking account of the need to progress environmental, social, and economic outcomes simultaneously;
- Use imagination, creativity, and innovation to provide products and services which maintain and enhance the quality of the environment and community while meeting financial objectives;
- Understand and secure stakeholder involvement in sustainable development; and
- Use resources efficiently and effectively.

Issues bearing upon the outcomes will include choice of material; energy efficiency; choice of subcontractors; messages transmitted during promotion; disposal at the end of the product's useful product life; product safety; and one's responsibility towards colleagues, customers, shareholders, society, and others.

The designer's challenges are becoming ever more complex – and the responses must be kept within the time, cost, and performance boundaries set by the project manager. This 'iron triangle' of project management never goes away. And all the time we must be ready to mediate conflicts between our obligations to the environment, to future generations, to customers, and to the wider business community. The Volkswagen Case (see Annex 3.1) shows that this balance is not easy to achieve when a business is striving to maintain a strong market position but loses sight of ethical priorities.

BEING PROFESSIONAL

It seems that the true professional may be distinguished by an accepted code of practice, a means by which an applicant can be assessed formally before being accepted into the profession, and a sense of belonging to, and wishing to serve, society. In this book, we have considered some of the ethical and statutory requirements which bear upon the engineer's professional work and must influence attitudes. In light of these pressures, what is the principal role of the professional engineer? How can it be learned? How should it be conducted? What does 'professional' mean?

In professional engineering practice, engineers must conform to design standards – in the selection of materials, for the specifics of stress, temperatures, tolerances, etc., know how to convert a customer requirement into a design specification, and then produce a design which satisfies this specification. But ethical issues and legislation impose disciplines which must be woven into the design process from the start. Principal criteria for professionals must include: expert and specialized knowledge in relevant disciplines and skills; high standards of professional ethics, behaviour, and work activities, since the duty owed to the client requires that the engineer will put the interest of the client ahead of his own interests; good work morale and motivation, having the desire to do a job well; and appropriate treatment of relationships with colleagues, with consideration shown to elderly, junior, or inexperienced colleagues, and those with special needs.

Section E3 of UK-SPEC requires the candidate to undertake engineering activities in a way that contributes to sustainable development; and section E5 demands that the engineer exercise responsibilities in an ethical manner. To be effective in either of these facets of the role implies prior thought and understanding both of the words on the page and their translation into action.

The professional engineer becomes aware through experience of many outside influences which impact on his/her work. The world is constantly changing; we must adapt, and always act according to the

DOI: 10.1201/9781003208433-8

appropriate Code of Ethics or Conduct. What if an explicit code of practice does not exist? What if it does exist but you disagree with it? What guidance can you rely on? Where can you go for advice? Worldwide the legislative and regulatory position is complex. How much can we and should we be aware of? Safety, quality, and all other supplementary requirements have a cost and a price. Who pays? How much is it worth paying? And who decides? The professional engineer must therefore:

- Read and understand all contractual and statutory requirements.
- Read and understand the required standards.
- Define and understand the terminology.
- Understand what guides and limits exist or may have to be applied.
- Reflect on the interfaces between personal responsibilities and those of customers, end users, the company, the team, and the general public.

Many engineers will feel deeply about the matters discussed in this book, and may demonstrate emotional reactions when major issues are brought to their attention. It is entirely legitimate at such times to show emotion, and we must not, indeed cannot, leave our emotions aside. Engineers are human too, and will occasionally want to display strong reactions; history demonstrates that it is easy, in attempting legitimately to oversee from above and impose statute or process, to ignore the human reactions of those affected by them. You cannot therefore rely upon rules and regulations – you must trust in the people around you and encourage open responses to a new situation (Figure 8.1).

We called up the notion of trust in Chapter 1, where we argued that the business's long-term view of the needs of the owner calls for confidence in a future, and hence for trust. Here we reach the same conclusion from a different perspective. The problem inherent in any system built on regulation rather than trust is that you cannot possibly monitor every stage of every process during the completion of a range of tasks. Tightening the rules simply drives trust out of the system – and drives self-confidence too. Rules should cover minimum standards and not try to achieve optimal ones – or no one will have any free space within which to exercise sound engineering judgement. In any potentially difficult situation we must trust our people; learn to trust even while following constraining legislation and regulations.

Figure 8.1 Emotions. *Printed with the permission of Glenys Tandberg 2021.*

This implies a readiness to allow empathy to be a part of our emotional response to those around us.

The greatest gift any engineer can develop is *imagination*, the asset from which great design and great leadership grow. That is why Henry David Thoreau wrote: 'If you have built castles in the air, your work need not be lost; that is where they should be - now put the foundations under them.' And as Mark Twain remarked: 'You cannot depend on your eyes when your imagination is out of focus.'

Perhaps in conclusion we should summarise the *four requirements* for every ethically professional engineer:

- Study the ethical codes of practice relevant in your context, *but be ready to move beyond them.*
- Learn from earlier practices in the organisation *and innovate from them.*
- Use your *imagination.*
- *Learn to take responsibility and to trust.*

Underlying these four simple pointers to behaviour we sense the need to be fully alert to the ethical pressures on us, to accept ownership, and to be ready to lead others to the same place.

FINALE

How does one attain truth and become wiser? Perhaps through the gradual acquisition of knowledge, appropriate skills, prudence, intelligence or intuition, and – ultimately – wisdom born of experience, generally in that order. It may be that learning the place of ethical thinking in our lives and careers requires a similar progression, and will not be acquired by reading.

We have reviewed in this short text the responsibility of engineers to be ethically aware as they engage in their work and plan their careers. It would be a bold engineer who would concede that he had behaved unethically; my own personal experience outlined in section 3 above shows how easy it can be to place one's priorities elsewhere and forget the essentials. Often the work context presents pressures which make the choices before the professional ever harder. Some may hope for release from their contextual challenges; the true professional will square up to them and do the best he can with the resources available, thinking on his feet and not being afraid either to make tough decisions or, when necessary, to seek advice. With time these decisions will become more challenging; projects will take place in a context of ever tougher markets, tighter budgets and timescales, and thus ever rising challenge. This need not be a frightening prospect; the best way to learn to deal with the world is to engage with it. John Keats wrote in a letter to a friend in 1818: 'I leaped headlong into the sea and thereby have become better acquainted with the soundings, the quicksands, and the rocks, than if I had stayed upon the green shore, and smoked a silly pipe, and took tea and comfortable advice.'

Readers of this text will have become aware that the works cited form only a very small proportion of the many books and journal papers on ethical thinking and its relevance to, and applications in, modern professions. Some readers, anxious for help in a confusing plethora of writing, may well be searching for explicit guidance in a personal dilemma. These searchers will probably be disappointed.

DOI: 10.1201/9781003208433-9

As has been said, challenging situations never repeat, and to the extent that common features arise the need for innovative thinking may trump our reach for standard solutions.

Against this background the need for innovation grows continually; the businesses, organisations, and indeed societies that survive will be those that refuse to believe the way things have been done so far is the only way. In the specific context of engineering, therefore, our profession can and must lead the way, and professional engineers are the ones who because of their wide awareness can seize the challenge. It is the theme of this essay that ethical standards across the board can and must rise to help us address the commercial challenges. Unless they do so, nothing else will win through in the long run.

If we are to behave ethically, where does one learn the appropriate behaviour? If at some stage in our lives awareness of ethical constraints became 'heart knowledge', how did this happen? And how is it that occasionally an engineer who would have doubtless wanted to describe himself as ethical participates in plainly unethical behaviours?

Such questions cannot be readily answered. We can only behave as ethically as we are aware at the time. Any account of the role of ethical attitudes in a professional context faces the massive issue that in the end, under guidance, we decide for ourselves. Recall the remark by Aristotle quoted at the beginning of this book; 'It is clear, then, ... that it is not possible to be good in the strict sense without practical wisdom, or practically wise without moral virtue.' Each of us must think afresh through each experience – and convert our experiences and thought into heart knowledge.

BIBLIOGRAPHY

The two great original thinkers on ethics are still worth reading. There are good modern editions of, and excerpts from, both thinkers. For Aristotelian Ethics, see Aristotle (2009), esp. Book 6, 'Intellectual Virtue', and Driver (2006). For Kantian Ethics, see, for example, Körner (1990) and Walsh (1975).

Many modern philosophers have written on ethical issues and may be read with benefit. Hare (1992) and Weil (1984) are selected here because many academic philosophical works tend to be fairly esoteric and most readers have time only for a small selection; both cover a wide field and can be used as a guide to further reading. For the engineering context, the two textbooks I have used in my teaching for many years are Velasquez (2012) and van de Poel and Royakkers (2011); both are available in e-book format, a feature which students may find useful. Driver (2006) is an excellent survey of the fundamentals of ethical theory. Seebauer and Barry (2001) is a good textbook for case studies. Chapter 4, especially, in Spier (2002) may be recommended.

WORKS CITED

Addagarrala, K., & Kinnicutt, P., 2018. Safety critical software ground rules. *International Journal of Engineering & Technology*, 7(2.28), pp. 344–350.

Aristotle, 2009. *The Nicomachian Ethics*. Oxford: Oxford World Classics (esp. Book 6, 'Intellectual Virtue').

Bazerman, M., & Tenbrunsel, A. E., 2011. Ethical breakdowns. *Harvard Business Review*, April.

Bishop, P, G., & Bloomfield, R. E., 1998. *A Methodology for Safety Case; Industrial Perspectives of Safety-Critical Systems*. Birmingham: Proceedings of the Safety-Critical Systems Symposium.

Bloomfield, R. E., & Bishop, P. G., 2010. *Safety and Assurance Cases: Past, Present and Possible Future*. Bristol: Safety Critical Systems Symposium.

DOI: 10.1201/9781003208433-10

Boldrin, M., & Levine, D. K., 2013. The case against patents. *Journal of Economic Perspectives*, 27(1), pp. 3–22.

Bowen, J., 2000. The ethics of safety-critical systems. *Communications of the ACM*, 43(4), pp. 91–97.

Bowen, J., & Stavridou, V., 1996. Safety-critical systems, formal methods and standards. *Software Engineering Journal (IEE/BCS)*, 8(4), pp. 189–209.

Bryson, J., 2018. Patency is not a virtue: The design of intelligent systems and systems of ethics. *Ethics and Information Technology*, 20, pp. 15–26.

BSI, 2016. *Ethical Design and Application of Robots*. London: British Standards Institution.

Committee on Science and Technology, 1986. *Investigation of the Challenger Accident*. House of Representatives, US Congress, pp. 1–256. Available at: https://digitalcommons.framingham.edu/.

Davis, M., 1991. Thinking like an engineer – The place of a code of ethics in the practice of a profession. *Philosophy & Public Affairs*, 20(2) (Spring 1991), pp. 150–167.

Davis, M., 2012. 'Ain't no one here but us social forces': Constructing the professional responsibility of engineers. *Science & Engineering Ethics*, 18, pp. 13–34.

Driver, J., 2006. *Ethics – The Fundamentals*, Oxford: Blackwell Publishing.

Dunn, C. a. B. B., 2006. A critique of Friedman 'the social responsibility of business is to increase its profits'. *Proceedings of the International Association for Business & Society*, 17, pp. 292–295.

EC, U., 2020. *United Kingdom Specification for Professional Engineering Competences (UKSPEC)*. 4th ed. London: Engineering Council UK.

Flores, A., 1998. The philosophical basis of engineering codes of ethics. In: *Engineering, Ethics and the Environment*, eds. Veselind, P.A. and Gunn, A.S. Cambridge: Cambridge University Press.

Friedman, M., 1970. The social responsibility of business is to increase its profits. *New York Times Magazine*, 13 September, Issue September 13.

De George R. T., 1981. Ethical responsibilities of engineers in large organisations. *Business and Professional Ethics Journal*, 1(1), pp. 1–14.

Hare, R. M., 1992. One philosopher's approach to business and professional ethics. *Business and Professional Ethics Journal*, 11(2), pp. 3–19.

Hasnas, J., 2008. *The Mirage of Product Safety*. Washington, DC: McDonough School of Business, Georgetown University.

HMSO, 1977. *The Patent Act*. [Online] Available at: www.legislation.gov.uk/ukpga/1977 [Accessed 2017].

HMSO, 2006. *Gowers Review of Intellectual Property*. London: HM Treasury.

IEEE, 1999. *Software Engineering Ethics and Professional Practices.* New York City, NY: IEEE Computer Society.

Klein, W., 2006. Business ethics from the internal point of view. *Journal of Business Ethica*, 64, pp. 57–67.

Kono, N., 2014. Brundtland Commission (World Commission on Environment and Development). In: *Encyclopedia of Quality of Life and Well-Being Research* ed. Michalos A.C. Dordrecht: Springer. https://doi.org/10.1007/978-94-007-0753-5_441.

Körner, S., 1990. *Kant.* 1st ed. London: Penguin Books.

List, J. A., & Momeni, F., 2017. *When Corporate Social Responsibility Backfires: Theory and Evidence from a Natural Field Experiment.* Cambridge, MA: National Bureau of Economic Research Working Paper 24169.

Manuele, F. A., 2010. *Acceptable Risk.* [Online] Available at: www.asse. org [Accessed 2018].

Nissenbaum, H., 1996. Accountability in a computerized society. *Science and Engineering Ethics*, 2, pp. 25–42.

Paté-Cornell, E., 2012. On 'Black Swans' and 'Perfect Storms': Risk analysis and management when statistics are not enough. *Risk Analysis*, 32(11), pp. 1823–1833.

Powers, C. W., & Vogel, D., 1980. The teaching of ethics in higher education. In: *Science, Technology, & Human Values.* vol. 5, No. 32 (Summer, 1980), pp. 29–32, Los Angeles, CA: Sage Publications, Inc.

RAE, 2009. *Autonomous Systems – Social Legal and Ethical Issues.* London: Royal Academy of Engineering.

RAE, 2011. *Engineering Ethics in Practice: A Guide for Engineers.* London: Royal Academy of Engineering.

RAE, 2014. *Statement of Ethical Principles.* London: Royal Academy of Engineering.

Schwab, B., 1996. A note on ethics and strategy – Do good ethics always make for good business. *Strategic Management Journal*, 17, pp. 499–500.

Seebauer, E. G., & Barry, R. L., 2001. *Fundamentals of Ethics for Scientists and Engineers.* New York, NY: Oxford University Press.

Spier, R. E., 2002. *Science and Technology Ethics.* London: Taylor & Francis.

Steen, M., & Van de Poel, I., 2012. Making values explicit during the design process. *IEEE Technology and Society Magazine*, Winter, 2012.

Sternberg, E., 1999. *The Stakeholder Concept – A Mistaken Doctrine.* Leeds: Centre for Business and Professional Ethics, University of Leeds.

Sternberg, E., 2000. *Just Business: Business Ethics in Action.* 2nd ed. Oxford: Oxford University Press.

UK Department for International Development, 2002. *Commission on Intellectual Property Rights – Integrating Intellectual Property Rights and Development Policy.* London: UK Department for International Development.

Van de Poel, I., & Royakkers, L, 2011. *Ethics, Technology & Engineering.* London: Wiley Blackwell.

Velasquez, M., Shanks, & Meyer, 1989. Calculating consequences: The utilitarian approach to ethics. *Issues in Ethics,* 2(1).

Velasquez, M. G., 1983. Why corporations are not morally responsible for anything they do. *Business & Professional Ethics Journal,* 2(3), pp. 1–18.

Velasquez, M. G., 2012. *Business Ethics – Concepts and Cases.* 7th ed. Pearson.

Vogel, P., 1980. *Ethics in the Education of Business Managers.* Hastings on Hudson: Institute of Society, Ethics & the Life Sciences (The Hastings Institute).

Walsh, W., 1970. Pride, shame and responsibilty. *The Philosophical Quarterly,* 20(78).

Walsh, W. H., 1975. *Kant's Criticism of Metaphysics.* 1st ed. Edinburgh: Edinburgh University Press, esp pp. 185–189.

Weil, V., 1984. The rise of engineering ethics. *Technology in Society,* 6, pp. 341–345.

Wolgast, E., 1992. *Ethics of an Artificial Person – Lost Responsibility in Professions and Organisations.* Stanford, CA: Stanford University Press.

ANNEX 1 – THE IET RULES OF CONDUCT

These principles, published by the Institution of Engineering & Technology (IET) in the UK, are typical of those established by many professional engineering bodies to guide their members.

1. The IET promotes and encourages ethical behaviour in the practice of science, engineering, and technology by all stakeholders. In so doing, the IET seeks to raise the level of public trust and confidence in the positive contribution to society made by science, engineering, and technology. Responsibility for professional and personal decisions and actions rests with the individual member.

2. Through its Knowledge Network and Rules of Conduct, the IET provides guidance and information sources to support members to take such decisions, and to act, ethically.

3. *Commitment to a shared code of conduct is a defining* characteristic of a profession. The IET Rules of Conduct are designed to guide members to meet the standard of professional conduct as specified in the Bye-laws.

4. The Rules of Conduct also aim to support members to take an ethical stance when balancing the often conflicting interests and demands of employers, society, and the environment. The IET aspires to promote the highest standards of conduct, and consequently its Rules are consistent with principles widely accepted amongst leading engineering bodies worldwide.

5. *In joining the IET, all members must agree to abide by the* Rules of Conduct. Members reaffirm their commitment to the Rules through the annual renewal of their membership. The Rules of Conduct are defined by members for members and are reviewed regularly to ensure they remain relevant.

Suggestions that a member's conduct has contravened the Rules are reviewed under the Institution's investigation and disciplinary procedures.

Extracts

1. These rules have been made in accordance with Bye-law 31. Unless a contrary intention appears, 'member' means a member of any category; and 'employer' includes 'client.' Except where inconsistent with the context, words implying the singular shall include the plural, and vice versa, and references to one gender shall include references to each other gender.
2. Members shall observe the provisions of the current Statement of Ethical Principles published by the Engineering Council and the Royal Academy of Engineering.
3. Members shall uphold the reputation and standing of the Institution.
4. Members shall observe the provisions of the Institution's Royal Charter and Bye-laws and any supporting regulations or rules.
5. Members shall keep their knowledge and skills up to date through planned professional development and seek to broaden and deepen that knowledge throughout their working life. Members shall keep adequate records of professional development undertaken. Members shall also encourage persons working under their supervision to do the same.
6. Members shall not undertake professional tasks and responsibilities that they are not reasonably competent to discharge.
7. Members shall accept personal responsibility for all work done by them or under their supervision or direction. Members shall also take all reasonable steps to ensure that persons working under their authority are both suitably equipped and competent to carry out the tasks assigned to them.
8. Members shall assess all relevant liability for work done by them or under their supervision, and, if appropriate, hold professional indemnity insurance.
9. Members whose professional advice is not accepted shall take all reasonable steps: (a) to ensure that the person overruling or neglecting that advice is aware of any danger or loss which may ensue; and (b) in appropriate cases, to inform that person's employers of the potential risks involved.

10. Members are expected to report to their employers any suspected wrongdoing or dangers they identify in connection with the member's professional activities. This includes: (a) any breach of professional obligations; and (b) bribery, fraud, or other criminal activity; miscarriages of justice; health and safety risks; damage to the environment; and any breach of legal obligations including any act of discrimination (in accordance with the Equality Act 2010).

11. Members shall support colleagues or others to whom they owe a duty of care who in good faith raise any concern about a danger, risk, malpractice, or wrongdoing which affects others.

12. Members shall neither advertise nor write articles (in any medium) for publication in any manner that is derogatory to the Institution or to the dignity of their profession. Neither shall they authorise any such advertisement or article to be written or published by others.

13. Members shall not recklessly or maliciously injure or attempt to injure, whether directly or indirectly, the professional reputation, prospects, or business of any other person.

14. Members shall at all times take all reasonable care to limit any danger of death, injury, or ill health to any person that may result from their work and the products of their work.

15. Members shall take all reasonable steps to avoid waste of natural resources, damage to the environment, and damage or destruction of man-made products. Lawful work undertaken by members in connection with equipment intended for the defence of a nation will not infringe this Rule 15 or Rule 14.

ANNEX 2 – UK-SPEC

This annex reprints extracts from the introductory section of the fourth edition of UK-SPEC, published by the Engineering Council (EC UK) in 2020.

THE PURPOSE OF UK-SPEC

This document is the UK Standard for Professional Engineering Competence and Commitment (UK-SPEC). The primary purpose of UK-SPEC is to explain the competence and commitment requirements that people must meet and demonstrate to be registered in each of the three registration categories: Engineering Technician (EngTech), Incorporated Engineer (IEng), and Chartered Engineer (CEng).

This document also explains why professional registration is important, how to achieve professional registration, and what engineers and technicians must do to maintain professional registration, including membership of a licensee organisation, the requirement to maintain and enhance competence, and the obligation to act with integrity and in the public interest.

WHO UK-SPEC IS FOR

Many different users will find this document useful. However, it has been written primarily for these audiences:

- Individuals who are thinking about becoming professionally registered
- Licensees and professional affiliates through which engineers and technicians become registered
- Employers of engineers and technicians
- People responsible for engineers' education or training

Professional registration verifies that an individual can meet the engineering and technological needs of today, while also anticipating the needs of, and impact on, future generations. Both in the UK and overseas, professional registration gives employers, government, and society confidence in the engineering industry. In this way, professional registration offers safeguarding assurances. Registration demonstrates that an engineer or technician has reached a set standard of knowledge, understanding, and occupational competence. It also demonstrates an individual's commitment to professional standards and to developing and enhancing through continuing professional development (CPD).

People who gain further qualifications or experience over the course of their careers can be assessed for another registration title. Many people continue to develop their competence to enable them to move from EngTech to IEng or CEng, or from IEng to CEng.

Professional registration sets individual engineers and technicians apart from those who are not registered. Gaining a professional title establishes a person's proven knowledge, understanding, and competence to a set standard and demonstrates their commitment to developing and enhancing competence. Registration increases a person's earning potential and establishes credibility with peers across the profession. The professional qualifications of EngTech, IEng, and CEng are internationally recognised.

Maintaining registration requires continued membership of a licensee. licensees, in turn, can help registrants find development opportunities through exposure to new developments, training, or networking opportunities. In addition, the criteria of the UK-SPEC provide a useful framework for CPD, particularly for engineers and technicians aiming for a professional registration title. Achievement of registration can demonstrate a person's readiness for promotion or help them secure new roles or contracts.

ANNEX 3 – SOME CASE STUDIES

Many good case studies may be found on related websites and in relevant texts; Seebauer and Barry (2001) is useful in this regard. Included here is a selection from the case studies which I have set for students in recent years, modified only where needed to make the phraseology suit the text of the present work. The challenges set for the students are also given. Students were required to conduct their own research into the background of the case and write 2500 words in response to the specific challenges I set. I always hoped to be taught something I did not know, and I was not often disappointed.

3.1 THE VOLKSWAGEN CASE – 'A TOLERANCE OF MISCONDUCT'

The problem with scandals of any kind is that they are rarely contained. Scandals spread because they are inherently centred on people. People are fallible, and this exacerbates the scandals they become caught up in. Political scandals are a good case in point. A leak occurs, followed by a witch hunt, followed at some point by an admission of the leak, then another witch hunt to decide whose head should roll, followed by the head of someone less important than the person whose head *should* have rolled. The fallout is often far more damaging than the initial leak itself.

Although neither political nor financial scandals are ever good for those concerned, they at least tend to have an end in resignation or imprisonment. The problem with corporate scandals, however, is that they often have no end, or if they do, the end does not come for some time. Although the media storm surrounding them may die down, the repercussions are felt for years to come.

Following the latest revelation of 'irregularities' in the amount of carbon dioxide pumped out by 800,000 of its vehicles, it now seems fair to say that this is likely to be a multi-year scandal for Volkswagen. Some in the car industry are looking at the CO_2 revelation with a pinch of salt, pointing to the fact that VW admitted this error rather than waiting for a US federal agency to detect it. Others have suggested that because it relates to only 800,000 cars – as opposed to the 11 million fitted with an 'emissions defeat device' – it is less important. But the scandal has now spread to focus on petrol engines as well as diesel; this should be of major concern not only for VW but for the rest of the industry. This has gone beyond specific cheat devices to encompass measurements with physical engines. A scandal is developing from which the company will recover only after many years.

Although the chief executive has resigned, and VW has shaken up the board and has also recruited an 'ethics expert', the company has done nothing yet to change the structure of the company – the two-board structure which has failed to govern the company is still in place – see diagram below. If VW had non-executive directors operating within a single board structure, as in the UK, it is perhaps more likely that questions about the sales of diesel cars might have been asked sooner.

Boards can only do so much, however, and the scandal appears to have been a crisis of management; merely to replace the former CEO Martin Winterkorn with Matthias Müller is not enough. Müller came from Porsche, part of the VW family, and is therefore an insider. In private, analysts suggest he is simply a placeholder leader, who will ride the storm for two to three years until a long-term successor can be identified. From some shareholders' perspective he is a safe pair of hands who can try to keep the company's reputation intact, while at the same time beginning the much needed clean-up.

But perhaps what VW needs is not a clean-up but a break-up. It needs a radical new chief executive who is a change from the past and can look at the company's assets dispassionately, to assess what should stay and what should go; just as importantly, a leader who can assess business practices and say what is valid and what is not.

British companies have a long history of recruiting 'axe men' external chief executives who are not afraid to make bold decisions. Some get it right first time – Bill Winters at Standard Chartered is a good example in this regard. But what VW could learn from British boardrooms is the realisation that scandals of this scale cannot be solved by looking inwardly. If the company, and more importantly its shareholders, wants to see real change and to see VW move beyond this enveloping scandal, it has to make radical moves.

We now learn that there has been a 'culture of tolerance of misconduct' in Volkswagen for many years (which did not, we are told, reach the boardroom). When engineers were unable to find a way to meet the tough pollution controls needed for the company to launch a sales campaign in the USA, a 'small group' decided to cheat. About 400 staff have had their digital devices seized and 87 employees subjected to 'extensive interviews.' The Chairman of the company, Hans Dieter Potsch, has said that: 'We will be relentless in seeking to establish who was responsible – rest assured they will be brought to account. We will investigate in all directions and with no taboos, everything is on the table, nothing will be swept under the carpet.' The incoming CEO said: 'There's a need for greater humility,' but he added, 'I don't think I'll be going down on my knees.' (James Quinn, writing in the *Daily Telegraph*, 4 November 2015, modified and augmented, with permission)

1. Where does a corporate culture, good or bad, spring from?
2. Does such a culture reach everywhere in an organisation? If the 'culture of tolerance of misconduct' did *not* reach the boardroom, is it possible that only part of the company was responsible?
3. If you were an engineer with VW and had been asked by your manager to participate in the design of a device to enable cars to bypass emission standards, or had known of such a proposal, how would you have reacted?
4. What should be the priorities of the incoming CEO?
5. If you were appointed to the board as 'ethics expert', what recommendations would you make?

3.2 THE MCDONNEL DOUGLAS DC-10 CARGO BAY DOORS

The McDonnell Douglas DC-10 was a three-engine wide-body airliner manufactured by McDonnell Douglas and first flown in 1970. It carried two turbofan engines mounted on underwing pylons and a third engine at the base of the vertical stabilizer; it was designed for medium- to long-haul flights and was capable of carrying a maximum of 380 passengers.

The US aircraft company Convair was responsible for the design and construction of the fuselage of the DC-10. As a subcontractor to McDonnell Douglas, the Convair Division of General Dynamics undertook detailed design work, but the design requirements and all major design choices were determined by McDonnell Douglas.

Most doors on an aircraft are of a 'plug' design; they open inwards, and are held in place in flight by the pressure difference between the inside and outside of the aircraft. However, the cargo bay door on the DC-10 opened outwards; the pressure difference in flight pushed the door open, so it was important to have a reliable locking mechanism.

Convair conducted a hazard analysis of the door, which postulated several scenarios where failure of the door could lead to loss of the aircraft. They also had good reasons to question the reliability of the locking mechanism as failures had occurred in both ground flight and trials.

Convair were limited in their ability to directly control the safety risk of an open cargo bay door for several reasons:

- They supplied safety analysis to McDonnell Douglas and were *contractually prohibited* from speaking directly to the regulator, the Federal Aviation Administration (FAA). Safety matters that Convair raised with McDonnell Douglas *were not always included* in documentation passed by McDonnell Douglas to the regulator, including the scenarios involving cargo bay doors opening in flight.
- The FAA were aware of the in-flight incidents, and elected *not* to issue an Airworthiness Directive. Instead they negotiated with McDonnell Douglas to issue a less enforceable Service Bulletin requiring minor change to the door design.
- It was unclear who would bear the cost of changes to the door design, particularly if those changes were made at the request of Convair rather than their customer, McDonnell Douglas.

The Director of Product Engineering at Convair issued a memo to his immediate supervisor. In unambiguous terms he challenged the safety of the cargo bay door and the adequacy of the changes made in response to the incidents. He predicted that at least one aircraft would be lost in-flight during the life of the DC-10.

This prediction was fulfilled on 3 March 1974 when Turkish Airlines Flight 981 crashed near Paris, killing all 346 people on board. The crash was caused when an improperly secured cargo door at the rear of the plane broke off, causing an explosive decompression which severed cables necessary to control the aircraft. At the time, it was the deadliest plane crash in aviation history; it remains the fourth-deadliest plane crash in aviation history and the deadliest involving a DC-10.

If you were the Director of Product Engineering, and your memo had been ignored:

1. What are your ethical responsibilities in the case, and to whom do you owe them? Are any of your responsibilities incompatible? (For example, it may be incompatible to share information about risk *and* to keep that information confidential.)
2. What features of the *organisational and contractual* context have made it difficult for you to meet your ethical responsibilities?
3. Would it be possible for Convair to meet their business objectives without restricting the free flow of safety information?
4. If the Director of Product Engineering leaked information about the locking mechanism to the press, would it be reasonable for Convair to take disciplinary action? Would this decision depend upon whether an accident had occurred? Give your reasons.
5. Should the Director of Product Engineering feel personally responsible for the accident? Give your reasons.

3.3 ETHICAL CONFLICTS FOR WESTERN FIRMS IN CHINA: THE CASE OF GOOGLE

Google Inc.'s stunning announcement that it might withdraw from China follows its long struggle with the ethical implications of doing business here, an endeavour which has forced it to make painful concessions to its public embrace of freedom of information.

Google entered the China market relatively late. It began offering a Chinese-language version of its search site in 2000, but didn't open offices in China until 2005 – years after rivals like Yahoo Inc. – a delay that gave local rival Baidu.com Inc. time to gain dominance in the market. Top Google executives were aware of the potential enormous opportunity in China, but feared that entering the market would conflict with their 'don't be evil' mantra and their often-stated goal of making the world's information 'universally accessible and useful.' In intense internal discussions that lasted more than a year, Google executives agonized over how to reconcile the perceived business need to enter China with its principles.

Other big technology companies had already drawn criticism for accommodating the Chinese government. Cisco Systems Inc. was criticized by free-speech advocates for selling China equipment that helps government censors to block websites. Rights activists and US politicians slammed Yahoo for helping Chinese police identify a Chinese journalist who allegedly used his Yahoo email account to relay to an overseas website the contents of a secret government order. The journalist was sentenced to ten years in prison.

In 2002, Time Warner Inc. decided to abandon a planned joint Chinese venture for its America Online division, with a top executive saying that it worried regulators would be able to demand copies of subscribers' emails.

But in the end, China's allure – the country already had more than 100 million internet users in 2005 – proved too great. In 2006, Google launched a special version of its website for mainland China, google.cn, whose results were scrubbed to remove links to sites the Chinese government found politically objectionable. "While removing search results is inconsistent with Google's mission, providing no information (or a heavily degraded user experience that amounts to no information) is more inconsistent with our mission," said a senior Google official at the time.

Many foreign companies operating in China have to grapple with ethical dilemmas, including how to deal with pervasive corruption and where to draw the line on cultivating the political connections that can help them navigate the country's complex regulatory terrain. Media companies, however, face even trickier decisions since they operate in an area that is the traditional preserve of the Chinese Communist Party and face political requirements that conflict with their core mission and values.

In recent years, China has sought to build a commercial media industry, realizing the importance of information to the national economy. At the same time, however, the Communist Party views the media as an important tool of political control – a pillar of the one-party state with a critical role to play in preserving its grip on power.

Google and other information companies, by operating in the domestic Chinese market, have to accept controls similar to those imposed on their domestic counterparts. Even though Google's Chinese website follows Chinese regulations, it too has been subject to crackdowns by authorities.

"This is the biggest and boldest statement than any large American company has made about this topic. The clarity with which they have said today that there is a limit with what they will put up with in China, or any other regime in the world, is breathtaking," said John Palfrey, a professor at Harvard Law School, who follows internet freedom issues.

Google, which hosts only a filtered version of its search engine inside China, took a different approach to other internet companies, to try to limit the potential damage of government control of users' information. Yahoo also hosted its e-mail servers in mainland China, which led to a case in 2005 in which Chinese journalist Shi Tao was sentenced to prison for ten years for releasing document to a democracy site after Yahoo China provided details about him to police.

Chinese internet analysts said they were shocked by the Google announcement, but at the same time they understood the reasons behind its threat to back out of China. They said the cyberattacks have exposed Google's inability to protect its users' privacy, which could have a huge impact on its commercial reputation since many Chinese regard Gmail as safe.

According to China IntelliConsulting Corp, there are around 30 million Gmail accounts in mainland China, and among Chinese internet users who use search engines – the vast majority of the more than total 338 million total – nearly 30% log on to Google at least once a week. The number of frequent Google users in China is around 40 million, IntelliConsulting estimates.

Lu Bowang, China IntelliConsulting's managing partner, says that although the actual number of Gmail account users within China is relatively small, they are 'all very active users of the internet, they have high demand for email stability and also very much rely on emails in their daily life.' If Google leaves China, its impact on the Chinese internet would be 'huge.' Mr Lu said Google's influence in China goes well beyond its role as a search engine. 'Simply put, the existence of Google in the Chinese market is a motivation for efforts by Chinese internet ventures to innovate. Without Google, the mechanism for such motivation will be gone as well.'

Wen Yunchao, a prominent Chinese blogger and free speech advocate, said that Google is giving the Chinese government 'a slap in the face. They refuse to sell out their principles and dignity for some petty profits. China's business environment may be worse off as a result, and its attractiveness to foreign investors may suffer', he said. Another blogger, Peter Guo, said that since last year the Chinese government has progressively tightened internet censorship and forced IT companies to cooperate. 'Google has been widely chosen as the major internet service provider by Chinese advocates of human rights, so Google is considered one of the greatest threats to the ruling CCP (Communist Party).' He said that after deviating from its 'don't be evil' principle and cooperating with censors, Google has realized that 'the Chinese government has a large appetite and that compromising is not always enough.' He added that 'I support the statement of Google from its headquarters which delivered an angry and uncompromised stand against the censorship of the Chinese government' (Jason Dean, *The Wall Street Journal*, 13 January 2010).

This case provides a valuable illustration of the ways in which conflict can arise when firms attempt to behave in an ethical manner but in doing so risk creating financial harm to shareholders or other groups of stakeholders. Studying the case may cause us to revise our view that maximisation of shareholder wealth is the best route towards ethical behaviour. We all know, and many admire, the role which Google plays in our interactions with the Internet; and we have all observed, and as customers been drawn into, the success story that is modern China. This case therefore has very specific resonances in the modern world; but none of us can stand on the sidelines when issues such as those in the case paper are discussed. If we find ourselves, at various points in our career, using our engineering and design skills in any commercial context we will probably face dilemmas and conflicts of one kind or another sooner or later.

Having familiarised yourself with the case, your task in this assignment is to address the following issues:

1. Google's actions cannot be said to have benefited their shareholders – but did they behave ethically? In fact, could a business decision which harmed shareholders' interests *ever* be said to be ethical?
2. It is often said that 'Good ethics make good business.' Discuss.
3. Put yourself in the position of a board member of a large multinational engineering company. Your decisions and actions may be affected by cultural and societal values in some of your markets. How do you address these issues?

3.4 THE MIRAGE OF PRODUCT SAFETY

By the time you have completed this short course you will have become familiar with some of the issues surrounding the concept of product safety, and the difficulties we face in our design tasks as we respond to the rising tide of safety regulation. This course has been about 'professional engineering practice', and reminds us that engineers must know how to convert a customer requirement into a design specification, and then produce a design which satisfies this specification, conforming in the process to many diverse standards. But legislation imposes disciplines too – and these must be woven into the design process from the start. Given the looseness with which the word 'safety' is used, how sure can we be that we have pinned the main issues down adequately for our purposes?

You will need to read: Hasnas, Prof. John (2008), 'The Mirage of Product Safety', McDonough School of Business, Georgetown University, Washington DC, USA; reproduced from *The Oxford Handbook of Business Ethics*, by kind permission of the author. This paper poses a series of challenges which may seem to undermine all our certainties. I have selected five:

1. Three views of the manufacturer's duty – the *contract* view, the *due care* view, and the *social costs* are presented in outline. In your opinion, which of these three views is the most correct – or the least incorrect – and why?
2. 'How much risk renders a product unsafe?' Examine your attitude to the question of the risk/benefit balance.
3. '... the most efficient way of preventing injuries is to place the burden of avoiding them on the party that can do so at the lowest cost.' Is this a sound method of attributing responsibility for safe design?
4. Where does responsibility for safety principally lie – with designers, management, contract lawyers, or the end user? What would you do to ensure your position as designer is well defined?
5. 'Because product safety is an inherently subjective concept, it ... is in the eye of the beholder.' Discuss, with special reference to products you have recently acquired in the open market.

3.5 THE FIRESTONE TYRE COMPANY: CORPORATE SAFETY POLICY IN PRACTICE

This case illustrates several features of product design and interaction which might lead us to suspect that, in today's connected business environment, absolute certainty on safety aspects of design may be hard to assure, and that research and risk assessments do not always provide the right answers – or even give appropriate guidance.

Radial tires were introduced to the US market by rivals Goodrich and Michelin in the late 1960s, and Firestone lacked one. The first radial tire developed and produced by Firestone was the ill-fated Firestone 500 Radial. Manufacturing of the new tire was performed on equipment designed to manufacture bias-ply tires.

During the 1970s, Firestone experienced major problems with the Firestone 500 radial. The Firestone 500 steel-belted radials began to show signs of separation of the tread at high speeds. While the cause was never proved, it is believed that the failure of bonding cements, used by Firestone to hold the tread to the tire carcass, may have allowed water to penetrate the tire, which in turn may have caused the internal steel wire to corrode. In March 1978, the US National Highway Traffic Safety Administration (NHTSA) announced publicly a formal investigation into defects of the Firestone 500. The NHTSA investigation found that the tread separation problem was most probably a design defect affecting all Firestone 500s.

In 1973, only two years after the 500's debut, Thomas A. Robertson, Firestone's director of development, wrote an internal memo stating: 'We are making an inferior quality radial tire which will subject us to belt-edge separation at high mileage.' Firestone introduced strict quality control measures in an attempt to fix the inherent problems; however, they were not successful in totally eliminating the basic faults. In 1977 a recall of 400,000 tires produced at the problematic Decatur plant was initiated. Firestone was considered to be less than co-operative with the NHTSA during the agency's investigation into the Firestone 500. Firestone blamed the problems on the consumer, stating under-inflation and poor maintenance.

On 20 October 1978, Firestone recalled over 7 million Firestone 500 tires, the largest tire recall to date. Congressional hearings into the 500 also took place in 1978. The tire was found to be defective and the cause of 34 deaths. In May 1980 after finding that they knew the tires were defective, the NHTSA fines Firestone $500,000 USD, which at that time was the largest fine imposed on any US corporation and the largest civil penalty imposed since passage of the 1966 National Traffic and Motor Vehicle Act. Multiple lawsuits were settled out of court and the constant negative publicity crippled the company's sales and share price. After years of bad publicity and millions paid out in compensation to victims, Firestone was losing vast amounts of money, and its name was severely damaged.

In 1996, several state agencies in Arizona began having major problems with Firestone tires on Ford Explorers. According to news reports, various agencies demanded new tires, and Firestone conducted an investigation of the complaints, tested the tires, and asserted that the tires had been abused or under-inflated.

On 6 September 2000, in a statement before the US Senate Appropriations Transportation Committee, the president of the consumer advocacy group Public Citizen, stated:

> There are a number of parallels between this recall in 2000 and the 1978 recall of the Firestone 500 . . . there was a documented cover-up by Firestone of the 500 defect, spurred by the lack of a Firestone replacement tire. When the cover-up was disclosed, the top management of the company was replaced as Firestone was severely damaged in reputation and economically. But a key difference is that the Firestone 500 was used on passenger cars, which rarely rolled over with tire failure. NHTSA documented 41 deaths with the 500, a recall, involving seven million tires.

The report went on to indicate that Ford also had a major role in the problems stating that

> The Ford Motor Company had instructed Firestone to add a nylon ply to the tires it manufactured in Venezuela for additional strength and that Ford had made suspension changes to the Explorer model available in Venezuela. Ford did not specify adding the nylon ply for U.S.-made Firestone tires nor did it change the Explorer suspension on US models at this time.

An abnormally high failure rate in Firestone's Wilderness AT, Firestone ATX, and ATX II tires resulted in multiple lawsuits, as well as an eventual mandatory recall. In 2001 Bridgestone/Firestone severed its ties to Ford citing a lack of trust. The lack of trust stemmed from concerns that Ford had not heeded warnings by Bridgestone/Firestone relating to the design of the Ford Explorer. In 2006, Firestone announced renewed efforts to recall tires of the same model recalled in 2000 after the tires were linked to recent deaths and injuries. The recall/replacement program was supported by a comprehensive advertising and consumer outreach campaign and over 6.3

million tires were replaced of the total 6.5 million affected. While the company believes that most of those tires unaccounted for have probably been scrapped long ago it is still trying to locate as many as possible.

The vice president of Quality Assurance for Bridgestone Firestone North America Tire LLC stated: 'Even though there are only a small percentage of these tires believed to be still in use, we are continuing to put safety first and are implementing this campaign to try and reach a group of consumers whose tires have not been recovered.'

In May 2000, the NHTSA contacted Ford and Firestone about the high incidence of tire failure on Ford Explorers, Mercury Mountaineers, and Mazda Navajos fitted with Firestone tires. Ford investigated and found that several models of 15-inch Firestone tires had very high failure rates, especially those made at Firestone's Decatur, Illinois plant. This was one of the leading factors to the closing of the Decatur plant.

The president of the public advocacy group Public Citizen stated before the Transportation Subcommittee of United States Senate Committee on Appropriations on 6 September 2000, that 'there was a documented cover-up by Ford and Firestone of the 500 defect.' The Executive Director for the Centre for Auto Safety in his statement before the Senate Committee on Commerce, Science and Transportation in Washington DC, 20 September 2000 stated: 'Emerging Information shows that both Ford and Firestone had early knowledge of tread separation in Firestone Tires fitted to Ford Explorer vehicles but at no point informed the NHTSA of their findings.'

The Ford Explorer was first offered for sale in March 1990. Ford internal documents show the company engineers recommended changes to the vehicle design after it rolled over in company tests prior to introduction, but other than a few minor changes, the suspension and track width were not changed. Instead, Ford, which sets the specifications for the manufacture of its tires, decided to remove air from the tires, lowering the recommended pressure to 26 psi. The maximum pressure stamped into the sidewall of the tire was 35 psi; however, tires should only be inflated to the pressure listed by the vehicle's manufacturer.

The failures all involved tread separation — the tread peeling off followed often by tire disintegration. If that happened, and the vehicle was running at speed, there was a high likelihood of the vehicle leaving the road and rolling over. Many rollovers cause serious injury and even death; it has been estimated that over 250 deaths and more than 3000 serious injuries resulted from these failures, with not all occurring on Ford Motor Company vehicles. It is estimated that 119 of the 250 deaths resulted from a crash with a Ford Motor Company vehicle.

Ford and Firestone have both blamed the other for the failures, which has led to the severing of relations between the two companies. Firestone has claimed that they have found no faults in design nor manufacture, and that failures have been caused by Ford's recommended tire pressure being too low and the Explorer's design. Ford, meanwhile, point out that Goodyear tires to the same specification have a spotless safety record when installed on the Explorer, although an extra liner was included into the Goodyear design after recommendations to that effect were made to Ford. Firestone included an extra liner in its product and this was then also used to replace tires on Ford Explorers.

A product recall was announced, allowing Explorer owners to change the affected tires for others. Many of the recalled tires had been manufactured during a period of strike at Firestone. A large number of lawsuits have been filed against both Ford and Firestone, some unsuccessful, some settled out of court, and a few successful. Lawyers for the plaintiffs have argued that both Ford and Firestone knew of the dangers but did nothing, and that specifically Ford knew that the Explorer was highly prone to rollovers. Ford denies these allegations. In a 2001 letter to Ford Motor Company Chief Executive, John T. Lampe, Chairman/CEO of Bridgestone/Firestone, announced that Bridgestone/Firestone would no longer enter into new contracts with Ford Motor Company, effectively ending a 100-year supply relationship.

1. The case paper uses phrases such as, 'while the cause was never proved, it is believed that . . .' and 'the . . . problem was most probably a design defect.' Is it inevitable that causes are indeterminate? Where does this leave the designer? The manufacturer? The end user?
2. How should responsibility (or blame?) be apportioned between the design teams and their managers?
3. How much responsibility should lie with Ford, the immediate customer, in the Explorer story? Justify your decision.

3.6 THE FORD PINTO CASE

In designing its 1971 subcompact model, the Pinto, the Ford Motor Company decided to place the gas tank behind the rear axle. Mounted in this position, the tank was punctured during tests by projecting bolts when hit from the rear at 20 mph. Ford decided, however, not to change the position of the gas tank, primarily in order not to increase redesign and production costs. An internal memorandum written in 1971 recommended that Ford should wait to do this until 1976 when the government was expected to introduce fuel tank standards (none existed in 1971). The memo estimated that by waiting until 1976 the Company could save 20.9 million dollars. In addition, the Company decided not to install a part costing $6.65 per unit which Ford engineers determined could reduce the risk of the gas tank puncturing in a collision at 20 MPH. In 1977 the National Highway Traffic and Safety Administration claimed that a safety defect existed in Pintos manufactured from 1971 to 1976. In 1978 Ford ordered a recall of nearly 2 million vehicles. Between 1971 and 1978, plaintiffs brought about 50 lawsuits against Ford in connection with Pinto rear-end collisions.

Did the Ford engineers involved in designing and testing the Pinto, who were aware of the problems associated with the placement of the gas tank behind the rear axle, fail to meet their ethical responsibilities by not taking steps to bring these problems to the attention of the public? Would they be likely to have been aware of an ethical perspective? Is it wrong to weigh the cost of a redesign against the value of human lives potentially saved? Is the public naive to expect the safety margins of a low-cost product to be the same as those of a Mercedes or a Rolls Royce?

The Ford Pinto Case is reported and discussed at length in DeGeorge (1981). See also Bazerman and Tenbrunsel (2011).

3.7 THE GOLDEN GATE BRIDGE SUICIDE BARRIER: THE ARGUMENTS FOR AND AGAINST

When the Golden Gate Bridge in San Francisco was finished in 1937 it was — with its span of 1280 meter (4200 feet) — the world's longest suspension bridge. It is considered to be one of the best and most beautiful examples of bridge design (see Figure 3). The American Society of Civil Engineers has included the Golden Gate Bridge in its enumeration of the seven wonders of the modern world. However, the bridge is also the US's most popular place to commit suicide; since

1937, at least 1300 people have killed themselves by jumping off the bridge, an average of 20 to 25 per year. Suicide prevention has in fact been a concern ever since the bridge was designed. Joseph Strauss, the chief engineer, is quoted as saying in 1936 (a year before the opening of the bridge):

> 'The guard rails are five feet and six inches high [i.e. about 1.7 m] and are so constructed that any persons on the pedestrian walk could not get a handhold to climb over them. The intricate telephone and patrol systems will operate so efficiently that anyone acting suspiciously would be immediately surrounded. Suicide from the bridge is neither possible nor probable.

However, for some reason, the height of the railing was reduced in the detail design to 1.2 meter (4 feet), so making it not too difficult to climb the railing and jump off the bridge.

Already in 1940, the Board of Directors discussed an 'anti-suicide screen' but decided against it on the basis of aesthetic and financial considerations. Another concern was that the screen might create dangerous wind resistance and make the bridge structurally unstable. Since then, proposals for a suicide barrier have been made every decade but were unsuccessful until recently. Below, we will consider some of the main arguments of the opponents (those who are against the installation of the barrier) and the proponents (those who are in favour of installation of the barrier).

Aesthetic Argument

Opponent: The Golden Gate Bridge has been praised for its transparency and openness. Any barrier design would destroy the view. Architect Jeffrey Heller, for example, stated: 'When you look straight out, you'll see through all this mesh, which will be sad enough, but looking straight down the roadway, it will become a cage. . . . That is far too high a price for our society to pay.'

Proponent: Aesthetic considerations can be accounted for in the design of the barrier.

Effectiveness Argument

Opponent: A suicide barrier will not be effective, since people who want to kill themselves will simply go somewhere else.

Proponent: Most suicidal people act on an impulse and, when prevented from actually killing themselves, they often seek help and loose the desire to die. According to a study by Dr Richard Seiden of UC Berkeley, the hypothesis that Golden Gate Bridge attempters will 'just go someplace else' is unsupported by the data. He studied 515 people who, from 1937 to 1971, were prevented from jumping from the bridge, and found that only 6% went on to kill themselves.

Economic Argument

Opponent: The barrier is not worth the costs. Even if the barrier would save 20 to 25 lives a year, it is not worth 50 million building costs plus the annual operation and maintenance costs.

Proponent: A human life cannot be expressed in monetary terms. Moreover, a cost-benefit analysis wrongly anonymizes the victims: 'You may not think that 24 lives a year is worth spending $50 million . . . well that is until it is your daughter, son, sister, brother, friend, husband, wife etc.'

Autonomy Argument

Opponent: One should not interfere with people's freedom to commit suicide: 'If I want to jump off the bridge, I don't think it is anyone's job to stop me.'

Proponent: The decision to kill oneself is usually made in an impulse. As one commenter puts it: 'I do believe that people have free will and that society should not interfere with their free will. But these people are mentally ill and their illness can be treated. But they cannot be treated if we let them jump off the bridge.'

Responsibility Argument

Opponent: It is not the responsibility of society or the directors of the bridge to try to prevent suicides. Moreover, it is wrong to use architecture to solve a social problem. In the words of Jeffrey Heller: 'You can't correct all of the sadness and evil in the world.'

Proponent: It is the responsibility of society and the bridge directors to act. As Jerome Motto, a past president of American Association of Suicidology, expressed it: 'If an instrument that's being used to bring about tragic deaths is under your control, you are morally compelled to prevent its misuse. A suicide barrier is a moral imperative. It's not about whether the suicide statistics would change, or the cost, or whether [it] . . . would be as beautiful . . . A barrier would say, 'Society is speaking, and we care about your life.' '

On 10 October 2008, the Board of Directors voted 14 to 1 to install a stainless-steel net which would be placed 6 meter (20 feet) below the deck and would collapse around anyone who jumped into it. The estimated costs are 50 million dollars. It is expected that construction of the net will take several years and will not start before a number of additional studies has been carried out.

Source: Accessed at The *Golden Gate Bridge Suicide Barrier* Case Study, from ENMG 504 at American University of Beirut. Cited also in van de Poel and Royakkers (2011).

3.8 THE SPACE-SHUTTLE CHALLENGER DISASTER, JANUARY 1986

On 28 January 1986, the Space Shuttle *Challenger* broke apart 73 seconds into its flight, killing all seven crew members aboard. The crew consisted of seven NASA astronauts – including two women (one of whom, high school teacher Christa McAuliffe, would have been the first teacher in space). The spacecraft disintegrated over the Atlantic Ocean, off the coast of Cape Canaveral, Florida. The disintegration of the vehicle began after a joint in its right solid rocket booster failed at lift-off. The failure was caused by the failure of O-ring seals used in the joint that were not designed to handle the unusually cold conditions that existed at this launch. The failure of the seal caused a breach in the solid rocket booster joint, allowing pressurized burning gas from within the solid rocket motor to reach the outside and impinge upon the adjacent attachment hardware and external fuel tank. This led to the separation of the right-hand aft field joint attachment and the structural failure of the external tank. Aerodynamic forces broke up the orbiter.

On 31 July the previous year, the following letter had been sent to senior engineering management at Morton Thiokol, the company leading the rocket booster programme:

MORTON THIOKOL INC.
Wasatch Division

Interoffice Memo

31 July 1985
2870:FY86:073

TO: R. K. Lund
                Vice President, Engineering

CC: B. C. Brinton, A. J. McDonald, L. B. Sayer, J. R. Kapp

FROM: R. M. Boisjoly
                Applied Mechanics - Ext. 3525

SUBJECT: SRM O-Ring Erosion/Potential Failure Criticality

This letter is written to insure that management is fully aware of the
seriousness of the current O-Ring erosion problem in the SRM joints from an
engineering standpoint.

The mistakenly accepted position on the joint problem was to fly without fear
of failure and to run a series of design evaluations which would ultimately
lead to a solution or at least a significant reduction of the erosion problem.
This position is now drastically changed as a result of the SRM 16A nozzle
joint erosion which eroded a secondary O-Ring with the primary O-Ring never
sealing.

If the same scenario should occur in a field joint (and it could), then it is
a jump ball as to the success or failure of the joint because the secondary
O-Ring cannot respond to the clevis opening rate and may not be capable of
pressurization. The result would be a catastrophe of the highest order -
loss of human life.

An unofficial team (a memo defining the team and its purpose was never
published) with leader was formed on 19 July 1985 and was tasked with solving
the problem for both the short and long term. This unofficial team is
essentially nonexistent at this time. In my opinion, the team must be
officially given the responsibility and the authority to execute the work
that needs to be done on a non-interference basis (full time assignment until
completed).

It is my honest and very real fear that if we do not take immediate action to
dedicate a team to solve the problem with the field joint having the number
one priority, then we stand in jeopardy of losing a flight along with all the
launch pad facilities.

R. M. Boisjoly

Concurred by:

J. R. Kapp, Manager
Applied Mechanics

[31st July 1985:

From: R M Boisjoly, Applied Mechanics

To: R K Lund, Vice-President, Engineering (and widely copied)

'*This letter is written to insure that management is fully aware of the seriousness of the current O-ring erosion problem in the SRM joints from an engineering standpoint.*

The mistakenly accepted position on the joint problem was to fly without fear of failure and to run a series of joint evaluations which would ultimately lead to a solution or at least a significant reduction of the erosion problem.....

If the same scenario should occur in a field joint (and it could) there is a risk of failure of the joint because the secondary O-ring cannot respond... may not be capable of pressurisation. The result would be a catastrophe of the highest order – loss of human life.

'*An unofficial team ... with leader was formed on 19 July 1985 and was tasked with solving the problem for both the long and the short term. This team is essentially non-existent at this time.*

It is my honest and very real fear that if we do not take immediate action to dedicate a team to solve the problem with the field joint is its number one priority then we stand in jeopardy of losing a flight'

(Signed) R M Boisjoly

(concurred by) ... Manager, Applied Mechanics]

Morton Thiokol's vice-chairman was told that 'this letter is a red flag.' NASA were alerted, but nothing was done. The night preceding the launch was very cold – NASA engineers later said they recalled having heard that it would not be safe to launch in very low temperatures. But NASA claimed later that the information they had been given did not provide sufficient grounds for them to declare the launch unsafe. The burden of proof was placed on those who were opposed to the launch – they were in effect required to prove that the flight would be unsafe. NASA was under great political pressure to maintain the programme timescale, following years of delay and overspend on the Shuttle programme.

Our wish to behave 'ethically' may require us to disagree with management. It may be wrong *not* to communicate our doubts – but we must be sure of our facts before putting a case to those who have the responsibility for the final decision. In doing so, it will be essential to have support and backing from others who share our views – but even without this we may have to act. It might be argued that Boisjoly should have 'blown the whistle' to alert other authorities of the danger. Students of this case have debated the issue since it first became a major topic for analysis by business schools. Every engineer must try to play the roles of Boisjoly, his manager in Morton Thiokol, and a senior NASA engineer, and ask: 'How would I have reacted if I had occupied this role?' It is not an adequate response to sit on the fence.

ANNEX 4 – INTELLECTUAL PROPERTY

The phrase 'Intellectual Property' (IP) refers to any asset owned by an organisation or an individual which is not on the balance sheet but resides in, and is demonstrated by, the knowledge and skills of the employees. Such assets, although real and potentially valuable, do not show on an organisation's balance sheet and cannot be readily valued. Management of balance sheet assets is of course a key element of financial oversight, an essential contributor to the success or failure of plans and forecasts. Similarly, off-balance sheet assets call for skilled management, all the more so as the future health of businesses comes to depend upon knowledge rather than traditional skills and markets. The rights attaching to the ownership of intellectual property of any kind are collectively known as IPR. Most nations have legislated to manage and protect IP generated within their borders, and it is in the context of these local statutes and the ethical issues they raise that engineers are likely to have cause to read and reflect.

There are several categories of IP:

Patents: A patent is an exclusive right granted to the inventor for a fixed period of time in exchange for the public disclosure of the details of an invention, method, process, or substance which is new, inventive, and industrially applicable. The geographical coverage of this protection is determined by the countries in which protection is granted; often a patentee will choose to register with several national patent offices to ensure wide coverage.

[Software may be patented only if, when run on a computer, the program produces a 'technical effect', generally an experienced improvement which must itself be in an area of patentable technology.]

Trademark: A trademark is a symbol or design (which may include a word or words) which instantly identifies the source of an object produced by an entity and distinguishing it from those of others.

Design rights: Designs may be protected in three different ways: copyright (see below), unregistered design rights, or registered designs. The actual details of design rights will vary depending on national law, and specialist advice may be needed.

Copyright: Copyright gives the creators of certain kinds of material the right to control the ways in which their material can be used. These rights start as soon as the material is recorded in writing or in any other way.

The latter two forms of IPR are automatically assigned; patents and trademarks must be applied for. Ownership of any of these rights will give protection against unauthorized use, copying, distribution, onward communication (by electronic transmission, rental, or lending to the public), and public performance.

The most recent UK legislation with regard to management of IP is the Patent Act (HMSO, 1977). A useful survey of IP is the Gower Review of IPR (HMSO, 2006), sponsored by HM Treasury, which surveys the whole field and makes 54 recommendations to improve policy and practice in the UK. It is not clear that any recent statute has been passed to adopt any of the recommendations in the report.

INDEX

Printed in the United States
by Baker & Taylor Publisher Services

Printed in the United States
by Baker & Taylor Publisher Services